ENCYCLOPEDIA OF PHYSICS

EDITED BY

S. FLÜGGE

VOLUME XLI/2

BETA DECAY

WITH 91 FIGURES

SPRINGER-VERLAG

BERLIN · GÖTTINGEN · HEIDELBERG

1962

HANDBUCH DER PHYSIK

HERAUSGEGEBEN VON

S. FLÜGGE

BAND XLI/2

BETAZERFALL

MIT 91 FIGUREN

SPRINGER-VERLAG
BERLIN · GÖTTINGEN · HEIDELBERG
1962

© by Springer-Verlag OHG / Berlin · Göttingen · Heidelberg 1962

Softcover reprint of the hardcover 1st edition 1962

Library of Congress-Catalog-Card Number A 56-2942

ISBN-13: 978-3-642-45983-2   e-ISBN-13: 978-3-642-45981-8
DOI: 10.1007/978-3-642-45981-8

Contents.

Experiments on β-Decay.

By

O. KOFOED-HANSEN and C. J. CHRISTENSEN.

With 91 Figures.

Introduction.

The investigations of β-decay of radioactive nuclei belong to the subjects in physics which are useful tools for the study of nuclear structure. Simultaneously, although to a much lesser degree, β-decay gives information about certain aspects of atomic physics. The description of β-decay classifies this type of transitions as a strictly quantum mechanical phenomenon. The fields responsible for the β-decay interaction do not have any classical analogue, and the properties of these fields can be studied only by measurements on β-decay processes. In this respect β-decay gives us basic physical information, and to a certain extent it may be said that apart from the classical electromagnetic fields β-decay is one of the aspects of field theory in general about which the most experimental material is available.

The experimental methods applied in β-decay investigations are very much the same as those used in the study of γ-rays and will not be repeated here. Descriptions of the techniques usually applied may be found in Vol. XXXI and Vol. XXXIII (especially the article on β-ray spectroscopes). Results from β-decay investigations with implications for nuclear structure may be found in XXXIX (transuranium elements.), Vol XL (nuclear reactions, levels and spectra) and in the article which describes the theory of β-decay in Vol. V, Part 2. The β-decay interaction has some important bearings on mesonic phenomena also, and these matters are dealt with in Vol. XLIII. In this respect β-decay has brought to our knowledge the existence of the neutrino and the lack of parity conservation in weak interactions.

Since so many aspects of β-decay have been dealt with in other volumes of this Encyclopedia, it is the intention in the present chapter to discuss only such matters as are of interest for the understanding of β-decay itself, i.e. for the experimental determination of the nature of β-decay and of the arbitrary constants characterizing the β-decay interaction, for the experimental verification of the theory of β-decay and for the measurements of the properties of neutrinos all of this including the parity experiments.

β-radioactivity finds numerous important technical applications. Again such applications are generally of the same nature as the applications of γ-rays and the technical methods are largely the same. Further it is felt that this subject falls outside the scope of the present chapter. Thus, due to the rather restricted nature of the present discussion of β-decay experiments many important papers on β-decay shall not be mentioned here although the information contained in such papers may be of the greatest importance for some of the subjects mentioned above and discussed in other volumes. In order to obtain a full survey of β-decay investigations the reader should also consult the above mentioned volumes.

A. General measurements.

1. Identity of β-particles and electrons. Fundamentals of nuclear β-decay. The present section contains some experimental results regarding the nature of β-particles emitted in radioactive decay. The origin of β-particles emitted from radioactive substances is partly *direct decay* of a *parent nucleus* into the *daughter or recoil nucleus* by the simultaneous emission of an *electron* and an *(anti-) neutrino*, and partly deexcitation of an excited nuclear state by electromagnetic interaction with the atomic electrons whereby the excitation energy is given off to an electron. The first process is called *nuclear β-decay*, the second process is called *internal conversion*. These two types of reactions give us the main sources of β-particles. Internal conversion is discussed in ALBURGER's article on nuclear isomers in Vol. XLII. In addition there are various higher order reactions like *internal pair creation* whereby nuclear excitation energy is given off directly to pair creation, *double β-decay* whereby two electrons are emitted simultaneously (see Sect. 27), emission of nuclear β-particles together with X-rays of continuous energy spectrum (see Sect. 28) etc. Unless special selection rules inhibit the ordinary decay the higher order processes may be neglected in the first approximation. When very precise measurements are performed the influence of the higher order effects should however be included as corrections. Associated with nuclear β-decay are phenomena like *β+-decay* whereby positively charged β-particles are emitted together with neutrinos, and *electron capture* whereby an orbital electron from, say, the K shell is absorbed in the nucleus and a neutrino is emitted. In the following we shall discuss these processes together with ordinary β-decay.

When nuclear β-decay occurs the *nuclear charge Z* is changed to $Z+1$ for the daughter and when *β+-decay* or *K-capture* occurs the nuclear charge of the daughter is $Z-1$. This has been verified by numerous experiments where the daughter is radioactive so that tracer chemistry can be used to identify the nuclear charge of the daughter. In this manner SODDY[1] recognized the nuclear origin of β-decay. Also other methods may be used for identification of the daughter, e.g. after K-capture the characteristic X-ray of the daughter atom are emitted. In this manner ALVAREZ[2] identified the X-rays of the daughter of K-capture as belonging to the nuclear charge $Z-1$ and thereby recognized the K-capture process.

The *sign of the charges* of β- and β+-particles is measured by magnetic deflection, e.g. as in all types of transverse magnetic spectrometers or in such longitudinal spectrometers where the direction of rotation of the beam around the spectrometer axis is observed. By such means it was early recognized that β--particles are negatively charged and β+-particles are positively charged.

The identity of the β-particle *charge* with one elementary charge unit has been shown directly by LADENBURG and BEERS[3]. Their instrument is illustrated in Fig. 1. A double β-ray spectrometer is used in the following way: A strong source of RaE is inserted in the source position and a fixed magnetic field H_0 is applied. The electrons focused on the exit slit are collected in a Faraday cup and the charge they give off to the cup is registered by means of a sensitive electrometer circuit. The field is then reversed and the ionization of the electrons in the ionization chamber is registered. In order to get to the Faraday cup a current sufficiently large that measurement can be performed with reasonable accuracy, so strong a source was used that the same number of β-particles could not be counted in an

---

[1] Cf. e.g., F. SODDY: The Chemistry of the Radio-Elements. London 1914.

[2] L. W. ALVAREZ: Phys. Rev. **54**, 486 (1938).

[3] R. LADENBURG and Y. BEERS: Phys. Rev. **58**, 757 (1940). — Y. BEERS: Phys. Rev. **63**, 77 (1943).

ordinary *GM* counter without appreciable loss, and therefore the source was left in position until it had decayed to a small fraction of the original intensity and the above measurements were then repeated with the Faraday cup replaced by a *GM* counter. The two ionization chamber currents, the current to the Faraday cup and the number of counts give the charge of the electrons as $-(4.80 \pm 0.03) \times 10^{-10}$ e.s.u. in good agreement with the usual value for $e = -(4.8028 \pm 0.0002) \times 10^{-10}$ e.s.u. [1].

The identity of the *charge-to-mass ratio* for β-particles with that for electrons has been demonstrated on β-particles by electric and magnetic deflection experiments[2] of the same type as those used for electrons. By such methods e/m values have been found in good agreement with the value found for electrons: $e/m = (5.27299 \pm 0.00016) \times 10^{17}$ e.s.u./g. By such experimental investigations speculations about the nature of the β-particles based on reported e/m values different from that for electrons and resulting from faulty technique have been eliminated.

The numerical identity of e/m for β⁻-particles and β⁺-particles has been demonstrated by methods similar to those by which the e/m value has been measured for β⁻-particles[3]. Recently PAGE, STEHLE and GUNST[4] have improved the method and demonstrated the relative e/m identity for β⁻-particles and β⁺-particles to within $\pm 7 \times 10^{-5}$.

The measurements of e/m for fast electrons have further-more demonstrated the *vari-*

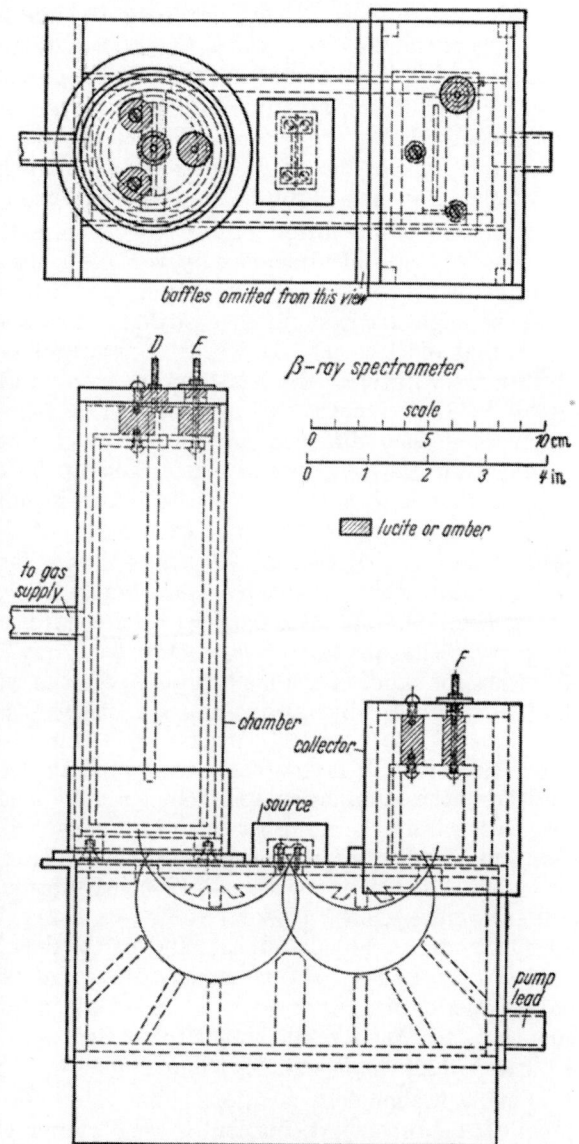

Fig. 1. The figure shows the double β-ray spectrometer used for direct e determination for β-particles from radioactive decay. (After Y. BEERS.)

[1] E. R. COHEN, J. W. M. DuMOND, T. W. LAYTON and J. S. ROLLETT: Rev. Mod. Phys. 27, 363 (1955). Cf. also COHEN and DuMOND's contribution to Vol. XXXV, this Encyclopedia.
[2] A. H. BUCHERER: Ann. d. Phys. 28, 513 (1909). — C. T. ZAHN and A. H. SPEES: Phys. Rev. 53, 357 (1938).
[3] A. H. SPEES and C. T. ZAHN: Phys. Rev. 58, 861 (1940).
[4] L. A. PAGE, P. STEHLE and S. B. GUNST: Phys. Rev. 90, 377 (1953).

ation of mass with velocity in agreement with the well-known formula from the special theory of relativity

$$m = \frac{m_0}{\sqrt{1 - \dfrac{v^2}{c^2}}}.$$ (1.1)

The presence of β^+-particles may be inferred from the measurement of the very characteristic *annihilation radiation* usually emitted when the β^+-particles are stopped in the material surrounding the β^+-source.

The annihilation radiation provides a means of measuring the *mass of β^+-particle* against that of the electron. The energy of the radiation has been measured[1]. The result is, that the relative difference in mass between the β^+-particle and the electron is less than $2 \cdot 10^{-4}$, within the experimental error.

Still all measurements of e and e/m cannot entirely kill the speculations along the lines of a possible difference between electrons and β-particles. Such measurements only demonstrate the identity inside the limits of experimental uncertainty and one might for example suggest that the *spin of β-particles* might be different from that of electrons. However, a beautiful proof of the actual identity of β-particles with electrons has been given by GOLDHABER and SCHARFF-GOLD-HABER[2]. They remark that the *Pauli principle* would not hold for a pair of particles if they differ in any property whatsoever. Consequently, when β-particles are slowed down in matter and finally captured in orbits around the nuclear charge, they should emit light corresponding to the transitions to lower levels, and if they are different from electrons they should emit characteristic X-rays since in that case the K-orbits are open for them. That this argument is valid is also demonstrated by the characteristic X-ray emission from mesonic atoms (Vol. XLIII). Here particles different from electrons show the postulated property. The absence of characteristic X-rays from β-ray absorption then furnishes the proof of the identity of β-particles with electrons. The experiment has been repeated by DAVIES and GRACE[3] who have shown that a fraction of less than 2.5×10^{-6} of all β-particles emitted from H^3 and absorbed in zirconium show zirconium X-rays. They used modern technique in their experiment, measuring the scintillation spectrometer pulse height distribution of the X-rays originating from the absorber (mainly bremsstrahlung from the stopping of the electrons). Their result is shown in Fig. 2. No trace of zirconium X-rays (16 kev) can be found. Also the application of H^3 constitutes an improvement because of the low maximum energy (18 kev) from this decay. This partly means that no energetic bremsstrahlung appears and partly that the slowing down time of the electrons is very low. DAVIES and GRACE estimate the slowing down time to be of the order of magnitude 10^{-14} sec and taking all due reservation they make the comment that their result shows that after $\sim 10^{-14}$ sec a fraction of less than 2.5×10^{-6} of all the β-particles from H^3 may be different from electrons.

Finally it should be mentioned that the occurrence of K-capture is another proof of the direct participation of usual atomic electrons in the β-decay interaction.

In order to make sure that the β^+-particle is the *antiparticle of the electron* one should mention again the presence of annihilation radiation occurring when β^+-particles pass through matter. Here, one may ask whether atomic electrons are the counterpart in the annihilation process. An argument pertaining to this point

[1] A. HEDGRAN and D. LIND: Ark. Fysik **5**, 177 (1952).

[2] M. GOLDHABER and G. SCHARFF-GOLDHABER: Phys. Rev. **73**, 1472 (1948).

[3] W. T. DAVIES and M. A. GRACE: Proc. Phys. Soc. Lond. A **64**, 846 (1951).

may be found in an experiment by KENDALL and DEUTSCH[1] who have measured the cross section for *annihilation-in-flight*. Their instrument is shown in Fig. 3. The β-ray spectrometer selects positrons in a certain energy interval. The positrons are focused on an Anthracene crystal of a scintillation counter, which is able to stop the positrons. This β-counter selects positrons which for some reason do not expend all their energy in the crystal, e.g. positrons which are annihilated in flight. The movable NaI(Tl) γ-scintillation counter selects γ-rays with an energy larger than $m_0 c^2 = 511$ kev. The β-counter and the γ-counter feed a coincidence unit. An output pulse from the coincidence indicates an annihilation-in-flight. With this instrument, both the angular distribution of annihilation gammas and the total cross section are measured. The results are (within experimental error:

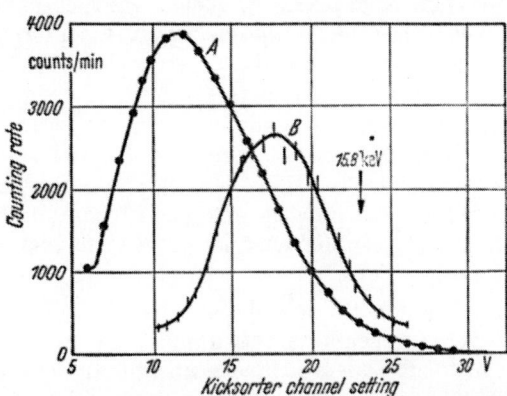

Fig. 2. Pulse height spectrum of X-rays from the slowing down of H³ β-particles in zirconium, curve A. Curve B is a calibration curve showing Bi L X-rays. (After W. T. DAVIES and M. A. GRACE.)

±5%) in good agreement with the theory based on the assumption of Z electrons per atom being able to annihilate with the positron.

Nuclear β⁻ and β⁺-decay leads to a *continuous energy spectrum* for the particles emitted. This was first shown for β⁻-decay by CHADWICK[2]. Before the

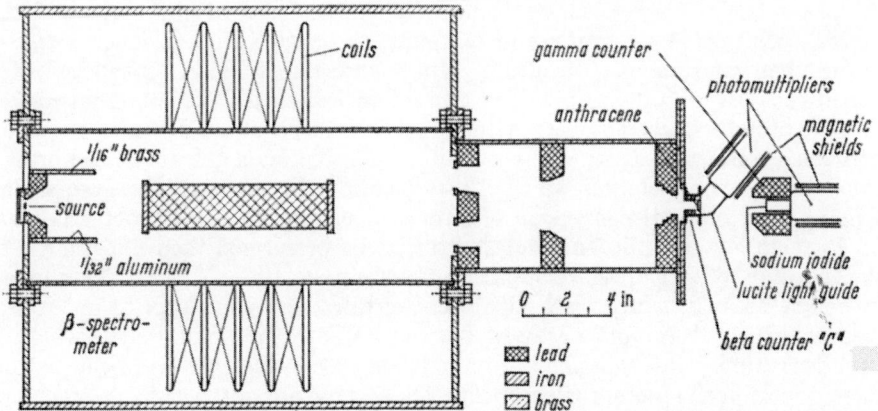

Fig. 3. The instrument used by KENDALL and DEUTSCH for measuring the cross section for annihilation-in-flight of positrons. (After H. W. KENDALL and M. DEUTSCH.)

neutrino hypothesis was introduced a great number of investigations were carried out in order to examine the possibility of the continuous spectrum being caused by some secondary effects. It was ruled out that β-decay should lead to a continuum of nuclear states in the daughter. This was demonstrated by the discrete γ-ray spectra following β-decay and the lack of sufficiently abundant γ-rays of continuous energy distribution following β-decay in radioactive series. Nor is the continuous spectrum caused by collisions of nuclear β-rays with atomic electrons either in the

[1] H. W. KENDALL and M. DEUTSCH: Phys. Rev. **101**, 20 (1956)
[2] J. CHADWICK: Verh. dtsch. Phys. Ges. **16**, 383 (1914).

decaying atom itself or in the layer carrying the source. This is most easily demonstrated by the lack of β^--particles accompanying β^+-decay. It was originally proved by a slightly different approach by Emeléus[1] who showed that the number of electrons emitted from a β-decaying sample very nearly equals the number of decaying nuclei. Further, calorimetric measurements (see Sect. 5) proved that the *average kinetic energy* of the emitted electrons equals the average value deduced from the spectrum and that no other relatively easily absorbed radiation is emitted with the residual energy. On the other hand the *maximum energy* of the β-spectrum and not the mean energy equals the energy balance in the β-decay process as derived from the mass difference between the nuclear states involved. This is very clearly demonstrated by the many cycles of nuclear reactions and β-decay energies involved in the mass determination of nuclei by means of reaction energies[2] and is further supported by electromagnetic mass determinations.

Occasionally speculations have occurred where it has been suggested that β-decay appears as the primary emission of some or other unknown particle which subsequently breaks up into an electron and a neutrino. Such a point of view has very little support in experiment or theory. As a heavy argument against such assumptions weighs the fact that β-spectra and neutrino recoil spectra show that the β-particle and the recoil participates in the *momentum and energy conservation* of a *direct three-body decay* (cf. Sect. 6 and 10). Whether or not a (heavy) meson participates in β-decay as virtual intermediate particle is a subject under discussion in the fundamental description of the theory of β-decay.

As a consequence of the above mentioned experimental results we shall in the following use as synonymous the words electrons and β^--particles, and likewise *positrons* and β^+-particles.

The internal conversion electrons mentioned above as the second phenomenon occurring in β-decay are of extranuclear origin. They produce a *discrete* rather than continuous β-spectrum. They arise due to electromagnetic interaction between orbital electrons and excited nuclear states. The internal conversion electrons usually compete with *γ-ray emission* and all properties and experiments concerning these electrons fall outside the main subject of the present chapter. For completeness we shall just mention the proof of the *extra-nuclear origin* of the internal conversion electrons. This experimental proof was given by Bainbridge, Goldhaber and Wilson[3] who performed their experiment on the internal conversion electrons from the isomeric state of Tc^{99}. This *isomer* (6 hours) decays mainly by a highly converted 2 kev transition to an excited state which by "prompt" decay of 140 kev leads to the ground state which in turn decays by β-ray emission with a half-life of 2.2×10^5 years and can be considered inert in the present experiment. The low energy of the isomeric transition means that the outer shells in the electronic structure participate to an appreciable extent in the conversion process. Consequently the chemical form of the Tc compound influences the halflife of the transition if the internal conversion is really an extranuclear effect. The small differences in half-life are measured by means of the *differential ionization chamber method* introduced by Ruther-

[1] K. G. Emeléus: Proc. Cambridge Phil. Soc. **22**, 400 (1924).

[2] For details compare the compilation of β-decay energies by R. W. King: Rev. Mod. Phys. **26**, 327 (1954) and by L. J. Lidofsky: Rev. Mod. Phys. **29**, 773 (1957) with the compilations of nuclear disintegration energies by D. M. van Patter and W. Whaling: Rev. Mod. Phys. **26**, 402 (1954); **29**, 757 (1957). Cf. also Li, Whaling, Fowler and Lauritsen: Phys. Rev. **83**, 512 (1951). — Li: Phys. Rev. **88**, 1038 (1952).

[3] K. T. Bainbridge, M. Goldhaber and E. D. Wilson: Phys. Rev. **90**, 430 (1953).

FORD[1]. The instrument is shown in Fig. 4. The sources 1 and 2 of approximately the same initial strength but different chemical form are placed in chambers A and B and the current to the common collector electrode i_{12} is measured when chamber A is so arranged that negative current flows here, while positive current flows to the collector in chamber B. The sources are then interchanged and i_{21} is measured under otherwise identical conditions. Finally the magnitude

$$\Delta i = (i_{12} - i_{21})/2$$

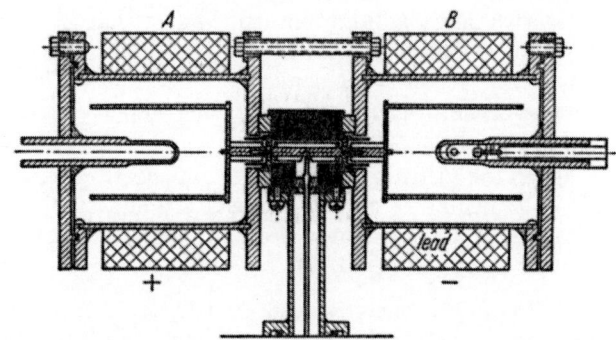

Fig. 4. Differential ionization chamber for the Tc compound half-life measurements. (After K. T. BAINBRIDGE, M. GOLDHABER and E. D. WILSON.)

is formed. Thereby any small difference in size and geometry of the two chambers is eliminated and Δi represents the difference in number of decays of the two sources, i.e.,

$$\exp(\lambda t)\, \Delta i = I_1 - I_2 + I_2 t\, \Delta\lambda \tag{1.2}$$

where λ is the decay constant of source number 1 and $\lambda + \Delta\lambda$ that of source number 2, and I_1 and I_2 are the respective intensities of the sources at $t = 0$. The results for Eq. (1.2) for sources 1 and 2 consisting both of Tc metal (compared as a test) and for $KTcO_4$ compared with Tc metal are shown in Fig. 5 as functions of time. The similar result for $KTcO_4$ compared with Tc_2S_7 which is not shown here gave as a result $\Delta\lambda/\lambda = (2.7 \pm 0.1) \times 10^{-3}$. Thus it can indeed be concluded that internal conversion is an extra-nuclear effect.

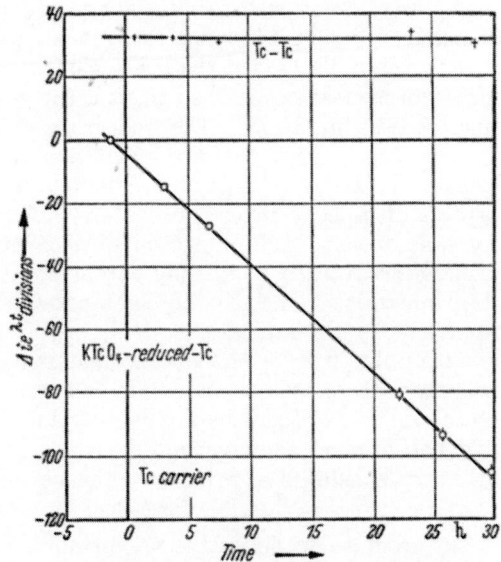

Fig. 5. The magnitude $\exp(\lambda t) \cdot \Delta i$ in arbitrary units as a function of time for Tc—Tc and $KTcO_4$—Tc sources. (After K. T. BAINBRIDGE, M. GOLDHABER and E. D. WILSON.)

To this information may be added the fact that internal conversion is accompanied by *characteristic X-ray emission* for the same Z as that in which the nuclear deexcitation occurs which shows that the nature of internal conversion is closely related to γ-ray emission. Conversely, this fact may be used to establish the Z of a daughter substance in β-decay, if this decay leads to an excited state in the daughter which undergoes decay by internal conversion. The energy differences between the K, L, M etc. shell conversion lines equal the energy differences between the corresponding absorption edges, and these energy differences

[1] E. RUTHERFORD: Wien. Ber. Series IIa. **120**, 303 (1911).

will, when carefully measured, also identify the daughter substance. It was actually by such methods that the first definite proof of the Z value of the daughter of β-decay was established by ELLIS and WOOSTER [1].

In this section we have given the clues to the nature of β-particles from radioactive decay and established the fact that they are identical with electrons. Furthermore we have discussed some of the differences between the nuclear β-rays and the internal conversion electrons. The investigation of the latter belongs to the discussion of electromagnetic deexcitation of nuclear states and shall not be discussed further.

When the identity of β-particles and electrons is stressed so carefully it should still be remembered that β-particles are emitted in states of unusual character as compared to accelerator beams of electrons, namely in practically *completely polarized states* (cf. Chap. D).

2. Maximum energies and half-lives. Theoretical notation. The energy spectrum of nuclear β-rays is continuous, as mentioned in the preceding section. The energy absorbed in a calorimeter with walls sufficiently heavy to stop all the β-particles equals the average energy of the energy distribution. But the energy release in the nuclear decay corresponds to the maximum energy of the spectrum. Consequently, energy has escaped observation. If energy conservation shall not be violated, we are forced to assume that energy is emitted in some form which is very difficult to observe. This assumption was first introduced by PAULI [2] at the Pasadena conference in 1931, and the theory of β-decay was developed along this line by FERMI [3]. The carrier of the energy that escapes ordinary detection devices is called a *neutrino*. One of the main problems for the experiments on β-decay is to investigate the properties of this particle and to check the assumptions underlying the general picture of β-decay according to this neutrino hypothesis. On the other hand β-decay is a useful tool for the investigation of nuclear properties just as is the case with γ-decay or any other form of nuclear transmutations. The main reason for this double nature of the aim of β-decay experiments lies in the fact that the interaction responsible for β-decay unlike the electromagnetic interaction, has no classical analog. These dual aspects of β-decay investigations are, however, tightly connected. In order to use β-decay as a tool for nuclear physics one has to rely on the β-decay theory. Using β-decay theory one may obtain the highest precision in maximum energy determinations and from the combination of lifetime measurements and maximum energy determinations (including branching ratios for complex decays) one may obtain values for nuclear matrix elements. On the other hand the same procedure that leads to maximum energy determinations gives the closest check on the validity of the theory of β-decay and the combination of maximum energies and half-lives is an important help in the experimental determination of the arbitrary constants entering into the perturbation term in the nuclear Hamiltonian which is responsible for β-decay.

In order to facilitate the understanding of this double purpose of β-decay experiments we shall first state the presently accepted status of β-decay theory, then use this theory in the description of maximum energy determinations and the determination of ft values from the combination of maximum energies and half-lives. In the following sections we shall then discuss the experiments on which the currently accepted β-decay theory is based.

[1] C. D. ELLIS and W. A. WOOSTER: Camb. Phil. Proc. **22**, 844 (1925).
[2] e.g. W. PAULI: Inst. Solvay, 7me Conseil, Bruxelles 1933, p. 324.
[3] E. FERMI: Z. Physik **88**, 161 (1934).

The *β-decay interaction* term of the Hamiltonian is to the first approximation found to be given by the following expression

$$H = H_V + H_A \qquad (2.1)$$

where H_V and H_A are the *vector* and *axial vector* interactions respectively given by

$$H_V = \Psi^* \, \Phi \psi^* \, (C_V + C_V' \gamma_5) \, \varphi - \Psi^* \vec{\alpha} \, \Phi \psi^* \vec{\alpha} \, (C_V + C_V' \gamma_5) \, \varphi + \text{c.c.} \qquad (2.2)$$

$$H_A = \Psi^* \vec{\sigma} \, \Phi \psi^* \vec{\sigma} \, (C_A + C_A' \gamma_5) \, \varphi - \Psi^* \gamma_5 \, \Phi \psi^* \gamma_5 (C_A + C_A' \gamma_5) \, \varphi + \text{c.c.} \qquad (2.3)$$

where Φ, φ describe the annihilation of a neutron and a neutrino, respectively, and Ψ^*, ψ^* describe the creation of a proton and an electron, respectively, γ_5, $\vec{\alpha}$ and $\vec{\sigma}$ describe Dirac matrices and are here used to give the combination of the four component spinor expressions Ψ^*, Φ, ψ^*, φ a scalar form plus a pseudo-scalar form where the latter is included in order to describe the *parity non conservation*. A priori the β-decay interaction is written not only as a sum of H_V and H_A, but also a *scalar* (S), a *tensor* (T) and a *pseudoscalar* (P) interaction are included[1]. The experimental evidence to be discussed later shows however that there is no compelling evidence to include more than the vector and axial vector interactions. The *selection rules* derived for the different coupling types make it possible to distinguish between the groups of couplings:

> *Fermi Couplings*   S, V
>
> *Gamow-Teller Couplings*   T, A

The β-decay probability $P(W) \, dW$ and the β-spectrum and angular correlations between the particles involved in the decay may be calculated from Eqs. (2.1) to (2.3). From general arguments about nuclear structure and from the form of these equations it follows that β-decay may be grouped according to the *multipole order* and *parity change* of the transition. Furthermore, it follows that only the largest term in the multipole expansion compatible with the *selection rules* on nuclear spin and parity in question need to be considered.

Then the β-spectrum, $P(W) \, dW$, of a definite transition may be written[2]

$$P_\pm(W) = \frac{1}{2\pi^3} F(\mp Z, W) \, p W (W_0 - W)^2 C_n \qquad (2.4)$$

where W is the total energy of the electron including the rest mass and connected with the kinetic energy E and the momentum p by

$$W = E + 1 = \sqrt{p^2 + 1} \qquad (2.5)$$

where we use the usual system of units applied in β-decay with $m_0 = c = \hbar = 1$. $F(\mp Z, W)$ is the *Coulomb correction* for allowed transitions given by

$$F(Z, W) = 2(1 + \gamma) \, (2p R)^{2(\gamma - 1)} \frac{|\Gamma(\gamma + i \, y)|^2 \exp(\pi \, y)}{[\Gamma(2\gamma + 1)]^2} \qquad (2.6)$$

where

$$\gamma = \sqrt{1 - (\alpha Z)^2}, \qquad (2.7)$$

$$y = \alpha Z W/p, \qquad (2.8)$$

[1] For details concerning the interaction term and its invariance properties see E. J. Konopinski: Annual Rev. Nucl. Sci. 9, 99 (1959) and the article on the theory of β-decay in Vol. V, Part 2.

[2] A. M. Smith: Phys. Rev. 82, 955 (1951). — D. L. Pursey: Phil. Mag. 42, 1193 (1951). — E. J. Konopinski and G. E. Uhlenbeck: Phys. Rev. 60, 308 (1941). — E. Greuling: Phys. Rev. 61, 568 (1942). — T.D. Lee and C.N. Yang: Phys. Rev. 104, 254 (1956).

and R is the nuclear radius. α is the finestructure constant and Γ the gamma function. The $+$-sign in $P_\pm(W)$ is applied for positron emission and the $-$-sign for electron emission. The factors C_n are called the *correction factors* for forbidden transitions and involve the nuclear charge Z, W, W_0 and the nuclear matrix elements in question together with the coupling constants. The expressions for C_n may be found in the papers quoted in connection with Eq. (2.4). To allow for the *parity not conserving term* in H we only have to insert e.g. $|C_A|^2 + |C'_A|^2$ for $|C_A|^2$ and $C_A C'_V + C_V C'_A +$ c.c. for $2C_A C_V$. The maximum energy of the transition W_0 is given by

$$W_0^- - 1 = E_{\beta^-}^{\max} = (M_{A,Z} - M_{A,Z+1}) \tag{2.9}$$

$$W_0^+ - 1 = E_{\beta^+}^{\max} = (M_{A,Z} - M_{A,Z-1}) - 2 \tag{2.10}$$

where $M_{A,Z}$ is the atomic mass of the parent nucleus and $M_{A,Z\pm1}$ that of the daughter.

The selection rules and matrix elements occurring in the different types of decay are given for reference in Table 1 in the usual notation introduced by KONOPINSKI. For each type of decay the correction factors C_n will contain as many terms as found from the number of absolute squares of matrix elements and cross terms between tensors of the same rank. Occasionally the terms may show the same behaviour as a function of W. This occurs e.g. in the case of allowed transitions where C is simply a constant given by

Table 1. *Selection rules and matrix elements in β-decay.*

Transition	Matrix elements	Tensor rank
Allowed $\Delta J = 0$ (no)	$\int 1$	0
	$\int \vec{\sigma}$ (no $0 \to 0$)	1
$\Delta J = 1$ (no)	$\int \vec{\sigma}$	1
1. forbidden $\Delta J = 0$ (yes)	$\int \vec{\sigma} \cdot \vec{r}$	0
	$\int \vec{r}, \int \vec{\alpha}, \int \vec{\sigma} \times \vec{r}$ (no $0 \to 0$)	1
	B_{ij} (no $0 \to 0$)	2
$\Delta J = 1$ (yes)	$\int \vec{r}, \int \vec{\alpha}, \int \vec{\sigma} \times \vec{r}$	1
	B_{ij} (no $0 \rightleftharpoons 1$)	2
$\Delta J = 2$ (yes)	B_{ij} (or S_{ij})	2
2. forbidden $\Delta J = 2$ (no)	A_{ij}, T_{ij}, R_{ij}	2
	S_{ijk} (no $0 \rightleftharpoons 2$)	3
$\Delta J = 3$ (no)	S_{ijk}	3

$$\left. \begin{aligned} C_0 = (C_V^2 + C_V'^2) \, |\int 1|^2 + \\ + (C_A^2 + C_A'^2) \, |\int \vec{\sigma}|^2. \end{aligned} \right\} \tag{2.11}$$

For simplicity we shall introduce the notation

$$\left. \begin{aligned} g_V^2 = C_V^2 + C_V'^2, \\ g_A^2 = C_A^2 + C_A'^2. \end{aligned} \right\} \tag{2.12}$$

It should be noted that we have chosen to include the absolute square of the coupling constant in C. In this respect we depart from the usual notation, but for all shape measurements or maximum energy determinations this is unessential since an arbitrary normalization is always introduced.

The total *decay rate* is found from Eq. (2.4) and is given by

$$\lambda = \int_1^{W_0} P(W) \, dW. \tag{2.13}$$

Here the integral

$$f = \int_1^{W_0} F(Z, W) \, p W (W_0 - W)^2 \, dW \tag{2.14}$$

is called the *Fermi integral*. Occasionally it is denoted f_0 in order to distinguish it from the integrals over the different shapes of forbidden transitions which are then labelled according to the type of transition in question. However, as may be seen from Table 1, most forbidden transitions include many different shape terms, and it is seen that only those decays for each order of forbiddenness which show the maximum spin change are associated with only one nuclear matrix element and thereby with a definite shape that can be calculated independently of our knowledge of nuclear structure. Such transitions are therefore called *unique shape transitions* and often a Fermi integral for these transitions is given by

$$f_n = \frac{f_0 \langle C_n \rangle_{Av}}{\left(g_A^2 \sum_{ij\ldots} |S_{ij}\ldots|^2\right)}. \quad (2.15)$$

The product of the half-life and the Fermi integral is referred to as the *comparative half-life* or more often the *ft value*. This *ft* value is connected with the nuclear matrix elements by the expression

$$f_0 t = 2\pi^3 \ln 2 / \langle C \rangle_{Av}. \quad (2.16)$$

A method for the rapid evaluation of the Fermi integral and *ft* values has been given by MOSZKOWSKI[1]. The value of $\log_{10} ft$ without the Coulomb correction is found from $E^{max} = E_0$ in Mev and from t

[1] S. A. MOSZKOWSKI: Phys. Rev. **82**, 35 (1951).

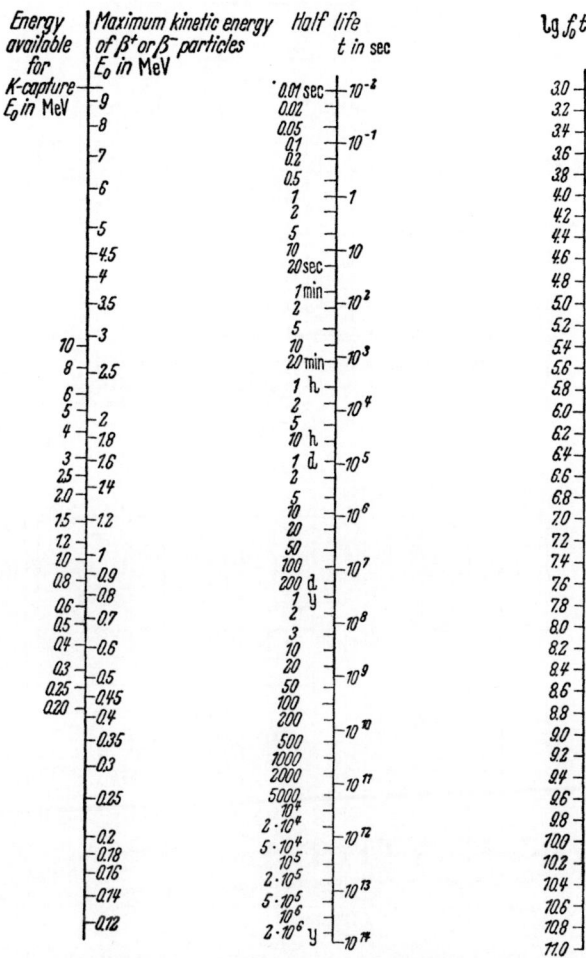

Fig. 6. Nomogram for obtaining $\log_{10} ft$ values from E_0 and t neglecting the Coulomb correction. (After S. A. MOSZKOWSKI.)

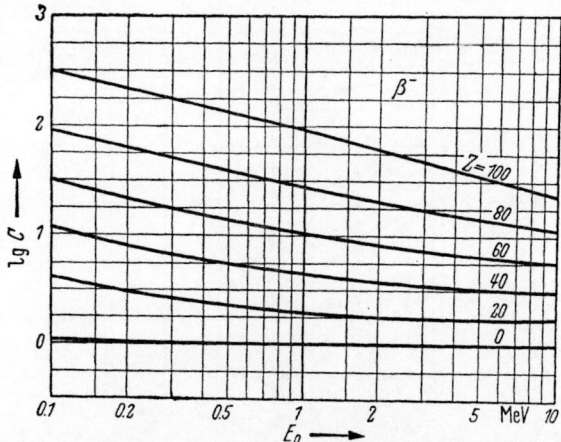

Fig. 7. Coulomb correction for electron emission $\log_{10} C$ to be applied to $\log_{10} ft$ from Fig. 6. (After S. A. MOSZKOWSKI.)

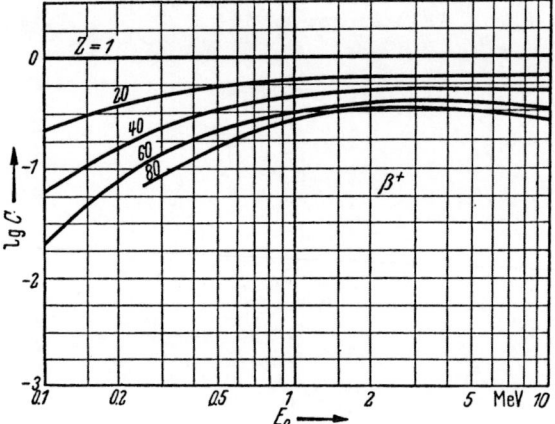

Fig. 8. Coulomb correction for positron emission $\log_{10} C$ to be applied to $\log_{10} ft$ from Fig. 6. (After S. A. Moszkowski.)

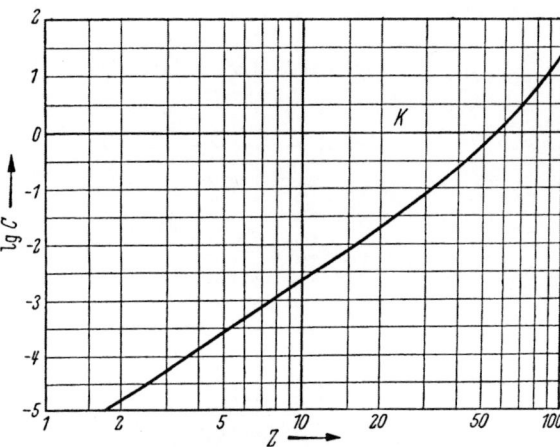

Fig. 9. Coulomb correction for K-capture $\log_{10} C$ to be applied to $\log_{10} ft$ from Fig. 6. (After S. A. Moszkowski.)

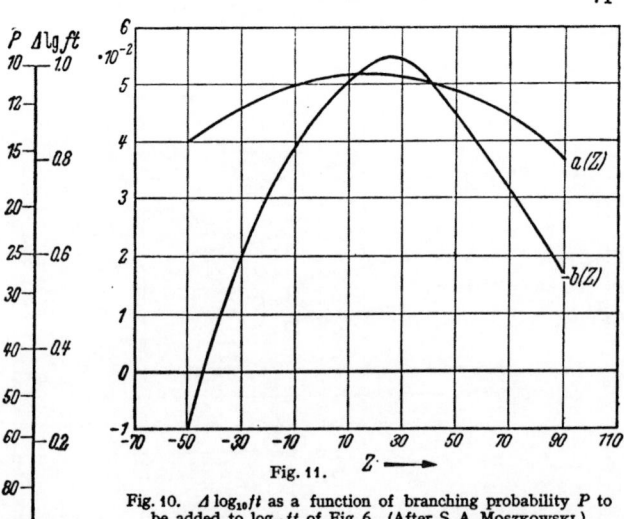

Fig. 10.

Fig. 10. $\Delta \log_{10} ft$ as a function of branching probability P to be added to $\log_{10} ft$ of Fig. 6. (After S. A. Moszkowski.)

Fig. 11. The coefficients $a(Z)$ and $-b(Z)$ used in connection with formula (2.17) for obtaining f_1. (After J. P. Davidson jr.)

from the nomogram shown in Fig. 6. To this is added the Coulomb correction $\log_{10} C$ found from Fig. 7 for electron emission, from Fig. 8 for positron emission and from Fig. 9 for K capture. Finally the correction for a possible branching ratio P is found from Fig. 10 and added. More accurate values have been given by Feenberg and Trigg [1]. Still more accurate values can be obtained by numerical integration of the actual spectrum shape. Numerical values for the allowed Fermi function [the Coulomb correction function, Eq. (2.6)] have been given by Rose, Dismuke, Perry and Bell [2] and by Stegun and Fano [2]. The energy dependent functions entering into the determination of the C_n correction factors have been tabulated by Rose, Perry and Dismuke [3].

It should be mentioned that f_1 defined by (2.14) has been estimated by Davidson jr. [4] who finds that

$$f_1 = f_0 \left[a(Z) (W_0^2 - 1) + \atop + b(Z) (W_0 - 1) \right] \quad (2.17)$$

where $a(Z)$ and $b(Z)$ may be found from the curves in Fig. 11.

[1] E. Feenberg and G. Trigg: Rev. Mod. Phys. 22, 399 (1950).

[2] N. M. Dismuke, M. E. Rose, C. L. Perry and P. R. Bell: Oak Ridge National Laboratory Report, No. 1222. — I. A. Stegun and U. Fano: Nat. Bur. of Stand. Appl. Math. Ser. 13.

[3] M. E. Rose, C. L. Perry and N. M. Dismuke: Oak Ridge National Laboratory Report, No. 1459.

[4] J. P. Davidson jr.: Phys. Rev. 82, 48 (1951).

If a detailed comparison of experimental data with theory is desired, the *atomic screening* should be taken into account. Corrections for this effect have been calculated by ROSE [1] (using the WKB approximation) and by REITZ [1] for a Thomas-Fermi model for the atom. The corrections usually amount to less than 3% except for the very heavy nuclei where the effect is quite important. Also *finite nuclear size* corrections should be considered at least for the heavier nuclei. This effect has been calculated by ROSE and HOLMES [2].

3. Maximum energies and half-lives. Experimental procedure. If the theoretical shape of the spectrum is known the entire spectrum may be used for the determination of the maximum energy. It was seen in the preceding section that the experimental β-spectra represent in general a mixture of several energy dependent terms each multiplied by an unknown matrix element. In other words, what energy dependence is to be expected is not a priori determined. Exceptions to this are the allowed transitions and the unique shape transitions. Even then, of course it is not known before measurements have been performed whether a certain spectrum is really allowed or first forbidden unique shape; this can indeed be decided only if additional information is available. For example if the spin and parity changes are known say to give an allowed transition, the theoretical expectation for the shape is known and the entire spectrum may then be used to determine the maximum energy. If this is not the case, the only proper procedure in obtaining the maximum energy is to measure the spectrum and find directly where the intensity of electrons drops to zero. Unfortunately due to the $(W_0 - W)^2$ factor the spectrum approaches the maximum energy with vanishing intensity and the method requires very difficult and at least tedious experimental work.

As mentioned above, if the spectrum shape to be expected is known on beforehand, the whole spectrum may be used to give the maximum energy. A convenient method for this procedure is the application of the so called *Fermi* or *Kurie* or *F-K plot* [3]. In this plot the quantity

$$\left(\frac{P(W)}{pWFC}\right)^{\frac{1}{2}} = \text{const}\,(W_0 - W) \tag{3.1}$$

is plotted against W where $P(W)$ is the observed energy spectrum. The result is a straight line cutting the W axis at W_0. Examples of Fermi plots are shown e.g. in Figs. 37, 38, 40 and 72. The plot obtained with $C = 1$ is called the *conventional Fermi plot*.

We now turn to the experimental procedure for obtaining Fermi plots. The raw data consist of the numbers of counts per unit time registered in a β-ray spectrometer. The β-ray spectrometer may be any one of the several types described in Vol. XXXIII. In most of the magnetic β-ray spectrometers used for the study of continuous spectra the number of counts is obtained as a function of the magnetic field H for fixed geometry, i.e. essentially for fixed *radius of curvature* ϱ of the electron trajectories and for fixed interval of acceptance of radii of curvature $\varDelta\varrho$. Since $p \sim H\varrho$ this means that the momentum spectrum is given by the number of counts per unit time divided by the interval size $\varDelta H \varrho = H\varDelta\varrho$ (apart from a normalization constant), i.e.,

$$P(p) = \text{const}\,N(H)/H \tag{3.2}$$

where this expression is evaluated for corresponding values of p and H. The energy spectrum is then

$$P(W) = P(p)\,dp/dW = P(p)\,W/p = P(p)/v \tag{3.3}$$

[1] M. E. ROSE: Phys. Rev. **49**, 727 (1936). — J. R. REITZ: Phys. Rev. **77**, 10 (1950).
[2] M. E. ROSE and D. K. HOLMES: Phys. Rev. **83**, 190 (1951) and Oak Ridge National Laboratory Report, No. 1022.
[3] F. N. D. KURIE, J. R. RICHARDSON and H. C. PAXTON: Phys. Rev. **49**, 368 (1936).

where W, p and v are corresponding values of energy, momentum and velocity of the electron. In proportional counter spectrometers (including scintillation counter spectrometers also) the directly observed distribution is the *pulse height spectrum*, $P(h)$. If pulse height, h, and energy are related by $h = h(W)$ (usually a linear expression $h \sim W$) then the energy spectrum is given by

$$P(W) = P\big(h(W)\big)\, dh/dW. \tag{3.4}$$

It should be noted here that in order to use the pulse height spectrum as directly proportional to the energy spectrum the linearity of $h(W)$ must be ascertained with such precision that dh/dW is constant inside the accuracy desired.

In order to proceed with the evaluation of the Fermi plot the quantities F and C, including the corrections mentioned in the preceding section, must be evaluated from the formulas and tables quoted, and then the magnitude of the function of Eq. (3.1) can be calculated and the Fermi plot drawn. The consistency of the data with Eq. (3.1) of course provides a check of the entire procedure. From the intersection of the Fermi plot with the W axis the maximum energy is obtained. The accuracy in careful work is of the order of 0.1 %. In such cases of course a correction for the experimental resolution must also be performed[1]

An upward deviation in the low energy part of the Fermi plot away from the straight line may indicate the presence of a *complex spectrum*. Such a spectrum is usually accompanied by internal conversion lines and/or γ-ray emission. The low energy components of the spectrum can therefore be investigated separately in a coincidence spectrometer where β-particles in the spectrometer in coincidence with γ-rays of the correct energy as measured in a scintillation spectrometer with pulse height selection are registered. A separate Fermi plot can then be obtained for such a spectrum and will yield a much more precise maximum energy determination than that which can be obtained by use of a subtraction procedure on the complex Fermi plot. If the source is too weak for coincidence measurements the latter method may of course be applied with appropriate caution. The methods sketched here are easily extended to more complex decay schemes, but of course the experimental difficulties increase with the complexity of the decay.

A convenient check on the observed maximum energies of complex decays is obtained by the measurement of the γ-ray energies in question and the construction of the entire decay scheme. The relative intensities of the different components of a complex β-spectrum and/or the relative intensities of the γ-rays will give information about *branching ratios*. As indicated in the preceding section, branching ratios must be known in order to obtain proper ft values.

Another important check on maximum β-decay energies is obtained from nuclear reaction data, especially Q *values*. Some of the most accurate maximum energies are indeed known mainly from Q values, and it may be said that few maximum energies had been determined accurately and carefully until Q value measurements had indicated that something was wrong with older β-spectroscopic data—often by amounts appreciable outside the limits of error quoted in β-spectroscopic work.

Lifetime measurements are usually carried out in the ordinary way by means of a counter and a stopwatch. There are no special precautions connected with half-life determinations of β-decay other than those applicable for α-decay or isomeric transitions. The same amount of care should be observed about chemical and isotopic purity. Effects from statistical uncertainties in the counting rates, background subtraction and overloading effects in the counters are taken into

---

[1] G. E. Owen and H. Primakoff: Phys. Rev. **74**, 1406 (1948).

account in the same manner as for other types of radioactive decay. In cases of a long series of radioactivities with successive nuclei of somewhat different half-lives special precautions are always necessary in order to obtain precise data. Just as for other types of radioactive decay some of the undesired daughter substances may be suppressed by energy selection of the particles investigated and, when lifetimes permit, chemical separations of the different types of radio-activity may be carried out. Isotopic separation is useful in the same manner. (It should be mentioned that when the radioactive atoms in an electromagnetic isotope separator are bombarded into a collector some of the very best sources for β-spectroscopic work are obtained.)

The large amount of valuable information obtained by β-spectroscopic work has been summarized in many different publications[1]. Also several tables of ft values have been given[2]. A histogram of the $\log_{10} ft$ values listed by KING[2]

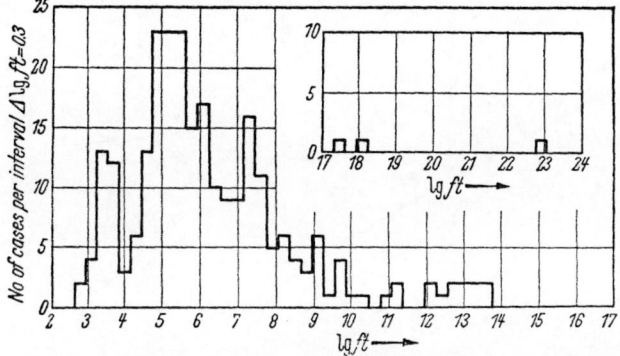

Fig. 12. Histogram of $\log_{10} ft$ values for which the spin assignments in the decay are reasonably certain.

for which the spin assignments are reasonably certain is shown in Fig. 12. This figure will be split into various components in Sect. C, but in order to illustrate the contents of this diagram it shall be mentioned that the first group of $\log_{10} ft$ values < 4.4 consists of the so called *superallowed transitions*. The following group starts out with *allowed unfavoured transitions* and mixes more and more with *first forbidden transitions. Second forbidden transitions* have $\log_{10} ft$ in the neighbourhood of 13. *Third forbidden* have $\log_{10} ft$ around 18.

The very definite importance of all this work accumulated in the indicated publications need hardly be discussed in detail here. As an illustration let us just mention that the ft values and the spin changes deduced from them were of very great help in establishing the nuclear shell model[3]. Also the presence of "magic numbers" among the nuclear neutron numbers N and proton numbers Z has been beautifully illustrated in plots of total β-energies for definite isotopic

[1] e.g. D. STROMINGER, J.M. HOLLANDER and G.T. SEABORG: Rev. Mod. Phys. **30**, 585 (1958). — K. WAY, L. FANO, M. R. SCOTT and K. THEW: Nuclear Data, NBS circular 499 with supplements. — K. WAY, R. W. KING, C. L. McGINNIS, M. WOOD and A. L. HANKINS: U.S.National Research Council: Nuclear Data Cards. — LANDOLT-BÖRNSTEIN: Zahlenwerte und Funktionen, Bd. 1, Teil 5. Berlin 1952 and New Series, Group I, Vol 1. — K. WAY, F. EVERLING, G.H. FULLER, N.B. GOVE, C.L. McGINNIS and R. NAKASIMA: U.S. National Research Council: Nuclear Data Sheets.

[2] A. M. FEINGOLD: Rev. Mod. Phys. **23**, 10 (1951). — R. W. KING, N. M. DISMUKE and K. WAY: Oak Ridge National Laboratory Report, No. 1450. — R. W. KING: Rev. Mod. Phys. **26**, 327 (1954). — L. J. LIDOFSKY: Rev. Mod. Phys. **29**, 773 (1957), and for K capture, J. K. MAJOR and L. C. BIEDENHARN: Rev. Mod. Phys. **26**, 321 (1954). — B. L. ROBINSON and R. W. FINK: Rev. Mod. Phys. **27**, 424 (1955); **32**, 117 (1960).

[3] M. G. MAYER, S. A. MOSZKOWSKI and L. W. NORDHEIM: Rev. Mod. Phys. **23**, 315 (1951). — L. W. NORDHEIM: Rev. Mod. Phys. **23**, 322 (1951).

spin transitions as a function of N.[1] The magic number $N = 50$ is illustrated in this way in Fig. 13 which is plotted according to the method used by Suess and Jensen from the data compiled by King [5].

At the same time such regularities as are revealed by these illustrative examples can of course in turn be used as checks on observed data.

In this chapter we have given a very short survey of maximum energies and ft values. We have included K capture data in the illustrations of the results

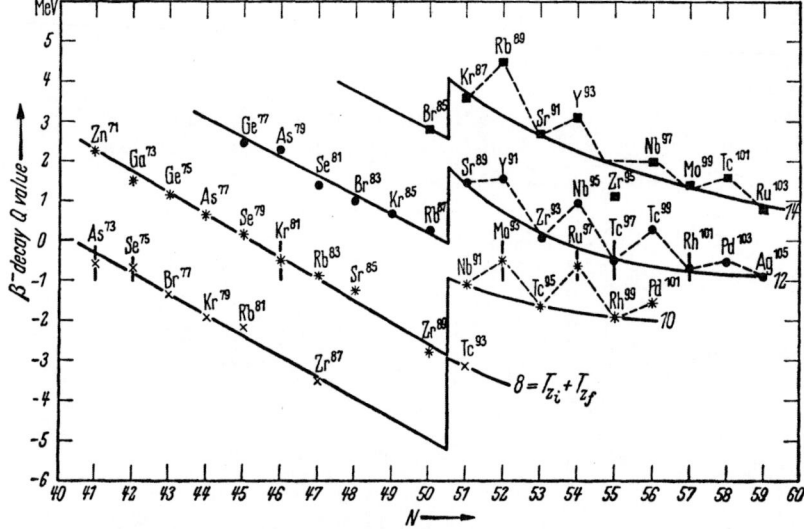

Fig. 13. β-decay Q values as a function of neutron number N. The curves connect transitions with the same sum of initial and final nuclear isotopic spin z-component. The decaying nucleus is indicated and the figure shows the effects of the magic number $N = 50$ on the decay energy and on the differences in pairing energies for neutrons and protons immediately after a magic number.

but further details about K capture, theoretical notation and energy determination have been postponed to Chap. F.

After these remarks about general measurements we now turn to the investigations of the details of β-decay.

B. Neutrino hunting.

4. Neutrino emission from reactions other than nuclear β-decay. When one wants to discuss the investigations in search of effects of neutrinos it is natural to start with those experiments which most clearly tell us that neutrinos have originated. Among such experiments are not only β-decay investigations but also a very large variety of meson experiments in which cloud chamber pictures or photographic emulsion techniques combined with energy and momentum conservation tell us that a light neutral particle has been emitted. In many of these cases it may further be concluded from the absence of showers in the neighbourhood of the event that the neutral particle cannot be a γ-ray. It is therefore natural to assume that the missing energy and momentum is carried away by a neutrino. This point is further stressed by the possibilities of theoretically accounting for the lifetimes and other decay features in question by means of the same ideas as those underlying the theory of β-decay. The best investigated cases of this type are the *π-meson decay*

$$\pi \to \mu + \nu; \tag{4.1}$$

[1] H. E. Suess and J. H. D. Jensen: Ark. Fysik **3**, 577 (1952).

the *μ-meson decay*

$$\mu \rightarrow \beta + \nu + \nu \qquad (4.2)$$

and *μ-meson capture*

$$\mu^- + p \rightarrow n + \nu. \qquad (4.3)$$

These reactions are discussed in Vol. XLIII, so we shall restrict the considerations here to nuclear β-decay.

5. Calorimetric mean energy measurements. The first β-decay experiments designed for the purpose of demonstrating that energy was being emitted in an unobservable way are the *calorimetric mean energy measurements* carried out by ELLIS and WOOSTER[1]. This type of experiments has since been carried out by many other experimentators[2]. In the experiments of ELLIS and WOOSTER the mean energy per disintegration of the β-rays from RaE($_{83}$Bi^{210}) was compared with the α-energy per disintegration of the daughter Po^{210}. RaE is especially well suited for this kind of work because it emits a continuous β-ray spectrum without any internal conversion lines and without any γ-rays. A source of RaE was inserted in one of the microcalorimeters shown in detail in Fig. 14a and shown inserted in the copper block A of Fig. 14b. The whole calorimeter system was furthermore surrounded by a wooden box lined with felt for protection against external heat disturbances. This latter box is not shown in the figure. Starting with a freshly prepared pure RaE source the heat developed by complete absorption of the RaE β-rays and the Po^{210} α-rays in a lead absorber in the microcalorimeter was measured as a function of time. A source is inserted in one of the microcalorimeters and a dummy in the other. Three minutes after inserting the source and dummy a temperature difference between the two calorimeters was registered by means of a thermocouple. This temperature difference was proportional to the heat given off by the source. Since only a comparison between the known α-energy and the mean β-energy is desired, no further calibration of the calorimeter had to be carried out apart from careful checks of the linearity of the relation between temperature difference and energy absorbed. Also, since Po^{210} is the daughter of RaE no absolute calibration of the source strength is needed. The results obtained by ELLIS and WOOSTER are shown in Fig. 15. Fig. 15a shows the total heating as a function of time in arbitrary units. The heating due to the Po^{210} is calculated from the asymptotic behaviour of the total heat curve and the known decay constants of RaE and Po^{210}. The subtraction of the heating due to Po^{210} from the total heating yields the result in Fig. 15b representing the heating due to RaE—in beautiful agreement with the known

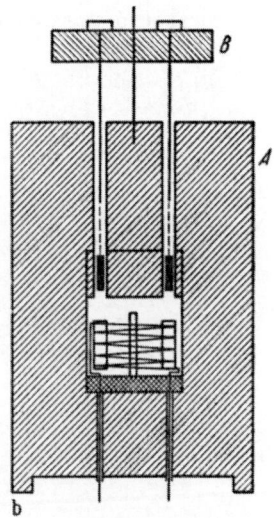
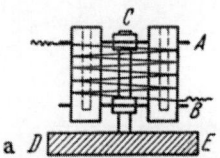

Fig. 14 a and b. The calorimeter used by ELLIS and WOOSTER. (After C. D. ELLIS and W. A. WOOSTER.)

[1] C. D. ELLIS and W. A. WOOSTER: Proc. Roy. Soc. Lond., Ser. A **117**, 109 (1927).

[2] L. MEITNER and W. ORTHMANN: Z. Physik **60**, 143 (1930). — I. ZLOTOWSKI: J. de Phys. 6, 242 (1935). — Phys. Rev. **60**, 483 (1941). — L. R. ZUMWALT, C. V. CANNON, G. H. JENKS, W. C. PEACOCK and L. M. GUNNING: Science, Lancaster, Pa. **107**, 47 (1948). — G. H. JENKS, F. H. SWEETON and J. A. GHORMLY: Phys. Rev. **80**, 990 (1950).

half-life of RaE. The ratio of heat per disintegration of RaE and Po[210] can be obtained from Fig. 15a, and the known α-energy of Po[210] then yields the mean energy for RaE $\langle E \rangle_{Av} = (350 \pm 40)$ kev. The later results of ZLOTOWSKI have given the improved accuracy $\langle E \rangle_{Av} = (320 \pm 5)$ kev. Both values are in very good agreement with the mean value calculated from observed spectral shapes: (317 ± 3) kev[1].

Since the early experiments of ELLIS and WOOSTER several improvements in the experimental technique have been obtained. Among these improvements we may mention the method of gas evolution for the measurement of heat in microcalorimetric work. In Fig. 16 the calorimeter developed by CANNON and JENKS is shown[2].

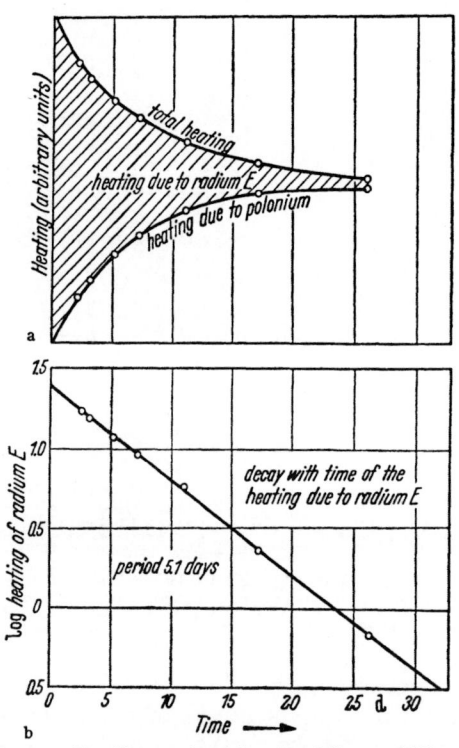

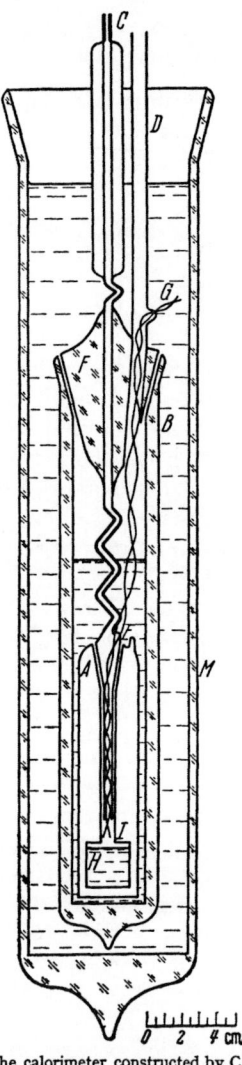

Fig. 15 a and b. The experimental results of ELLIS and WOOSTER. (After C. D. ELLIS and W. A. WOOSTER.)

Fig. 16. The calorimeter constructed by CANNON and JENKS. (After C. V. CANNON and G. H. JENKS.)

The Dewar flasks A, B and M contain the calorimeter A, the calorimeter bath B and the protection bath M. The inner systems are closed by ground glass joints E and F. Electrical leads are introduced through G and E so that a resistance heater can be placed in the calorimeter H for calibration purposes. The heat is measured by means of the gas produced by evaporation when the heat acts on liquid air in the calorimeter. The gas pressure over the calorimeter bath is controlled through the tube D by means of an auxiliary system not shown in the figure. A similar system and a device for the measurement of the

[1] For good RaE β-spectra see the discussion of RaE in Sect. 11.
[2] C. V. CANNON and G. H. JENKS: Rev. Sci. Instrum. 21, 236 (1950).

gas evolved is connected to the calorimeter proper through C. The system is usually operated at a pressure of a few cm Hg below atmospheric pressure.

By means of this instrument a precise determination of the average energy of H^3 β-rays has been performed by JENKS, GHORMLY and SWEETON[1]. Their result is $\langle E \rangle_{Av} = (5.64 \pm 0.04)$ kev. H^3 is one of the superallowed transitions. That means that the shape of the spectrum to be expected is given by Eq. (2.4) with C_0 an uninteresting constant. Consequently, if one relies on the theory, the measured average energy can be used to give a determination of the maximum energy. The result is (18.6 ± 0.2) kev in very good agreement with the values obtained spectroscopically[2].

Many other average energy determinations have been performed by means of calorimetric technic or by obtaining averages from β-spectroscopic investigations. The present two are shown only as illustrations of methods which are entirely different from the β-ray energy determinations carried out by means of electromagnetic spectroscopes. The results clearly show that energy is disappearing in some manner which is difficult to detect and is thus a support for the neutrino hypothesis. The agreement between the H^3 maximum energy found directly and that found from the mean energy assuming the theoretical energy distribution, is a good support for the theory itself. It should also be noted that the two experiments discussed here are somewhat different in method and rather different as regards the maximum energy of the β-ray employed.

6. Neutrino recoil experiments. Recoil experiments on neutrino emission have been described in many review articles[3] and the reader is referred to such articles for the older literature. The purpose of recoil investigations has so far been threefold. 1. The demonstration that the neutrino takes away not only energy, but also the proper amount of momentum is felt as a valuable clue to the existence of neutrinos. 2. The shape of the β-spectrum of allowed transitions is insensitive to the type of coupling in β-decay apart from the effects of the FIERZ[4] interference terms, whereas the absolute squares of the coupling constants $(|C_i|^2 + |C_i'|^2)$ are easily extracted from the recoil directional correlation[5] or the recoil energy spectrum[6] and additional information regarding the relative phases of the C's and C''s is gained from recoil directional-polarization-correlations [e.g. the neutron decay (cf. Sect. 17)]. Thus recoil experiments provide a tool for the experimental determination of the coupling constants in β-decay. 3. A series of secondary effects connected with the excitation of the atomic electrons during β-decay may be studied by recoil investigations.

In this section the classical recoil experiments (directional correlations and recoil energy spectra) other than the neutron decay will be discussed. A detailed examination of the neutron decay including polarization effects will be presented in Sect. 17. Finally the discussion of the extraction of coupling constants will be postponed to Chap. E. Before we discuss these three types of investigations

---

[1] G. H. JENKS, J. A. GHORMLY and F. H. SWEETON: Phys. Rev. **80**, 990 (1950).

[2] The H^3 spectrum is shown in Fig. 40.

[3] B. PONTECORVO: Rep. Progr. Phys. **11**, 32 (1946/47). — J. S. ALLEN: Amer. J. Phys. **16**, 451 (1948). — H. R. CRANE: Rev. Mod. Phys. **20**, 278 (1948). — C. W. SHERWIN: Nucleonics 2, No. 5, 16 (1948). — O. KOFOED-HANSEN: Physica, Haag **18**, 1287 (1952). — O. KOFOED-HANSEN [2], p. 357.

[4] M. FIERZ: Z. Physik **104**, 553 (1937).

[5] F. BLOCH and C. MØLLER: Nature, Lond. **136**, 911 (1935). — D. R. HAMILTON: Phys. Rev. **71**, 456 (1947). — E. GREULING and M. L. MEEKS: Phys. Rev. **82**, 531 (1951). — H. L. REYNOLDS, L. C. BIEDENHARN and D. B. BEARD: Oak Ridge National Laboratory Report, No. 1444 and Erratum. — O. KOFOED-HANSEN: Dan. Mat. Fys. Medd. **28**, No. 9 (1954).

[6] O. KOFOED-HANSEN: Phys. Rev. **74**, 1785 (1948).

separately we shall discuss the general difficulties underlying all three types of measurements.

The energy of recoil particles is generally very low. The kinetic energy E_R of a recoil nucleus of atomic mass number A after neutrino emission (K capture) of energy W_K is given by

$$E_R \,(\text{ev}) = (534/A)\,W_K^2 \quad (W_K \text{ in Mev}) \tag{6.1}$$

and the maximum kinetic energy of the recoil after β-decay is given by

$$E_R^{\max}\,(\text{ev}) = (534/A)\,(E^{\max} + 1.022)\,E^{\max} \quad (E^{\max} \text{ in Mev}). \tag{6.2}$$

This results in energies from a fraction of an electron volt up to $\sim$1000 ev in a few cases. The handling of such low energy ions presents quite a few problems in itself.

When the parent nuclei form parts of molecules or are deposited on a surface, essentially two types of difficulties appear. Firstly, the momentum of the recoil is more or less shared with other particles. Close collisions with the atoms of the molecule or the surface atoms are very probable shortly after the decay has taken place and during the breakup of the molecule and the escape from the surface. Thus the momentum distribution may be distorted to an appreciable extent. This is demonstrated very clearly in the experiments of Davis jr.[1]. The measurements were carried out on the recoils from the K capture decay of Be[7] with the Be[7] atoms deposited on tungsten. The results are shown in Fig. 17 where they are compared with the theoretically expected resolution curve (dotted line) for the electrostatic spectrometer used in the experiments calculated on the basis of the decay

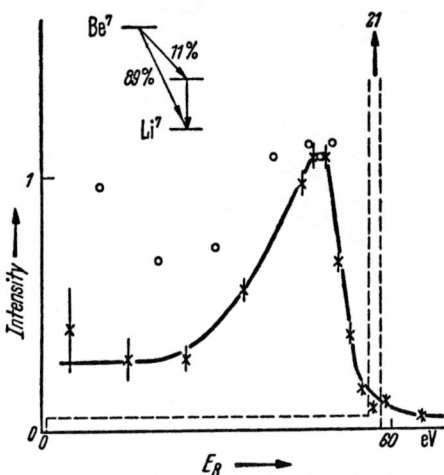

Fig. 17. The Be[7] recoil spectrum observed by Davis jr. compared with the experimental resolution to be expected if no surface effects were present. (After O. Kofoed-Hansen.)

scheme shown and neglecting surface effects. The experimental curve is smeared out towards the lower energies, and Davis jr. also showed that the surface undergoes severe changes with time as indicated by the circular points in Fig. 17. These points were obtained some hours later than the points indicated with crosses and which were obtained with a fresh surface. Secondly, when a molecule breaks up it is a very difficult job to estimate which piece of the molecule will get the electric charge and thereby subsequently be detected as the "recoil". This effect can of course entirely distort the recoil angular correlation or the energy distribution.

Recently it has been tried to overcome this difficulty in the decays of Li[8] and Na[24]. These experiments will be discussed later in this section.

Fortunately, there exist a few monatomic gases with desirable properties for recoil investigations. When experiments are carried out on such radioactivities, none of the above mentioned difficulties appear. On the other hand, the choice of the dimensions of the recoil measuring instruments becomes somewhat more complicated because the source has a tendency to spread out over the entire

[1] R. Davis jr.: Phys. Rev. 86, 976 (1952).

vacuum system of the instrument. And, of course, also a certain amount of surface effects are to be expected when the recoils hit the collector. At any rate the most reliable recoil experiments have been performed on inert radio-active gases and we shall confine our discussion to such experiments, thereby illustrating some of the tricks employed in the development of suitable neutrino recoil spectrometers. Finally, we may say that when the behaviour of β-recoils from inert gases is known, experiments on bound parent nuclei can undoubtedly be used for the study of many interesting atomic effects connected with molecular break-up after radioactive decay of one of the atoms in the molecule. For the present discussion, however, we shall mainly discuss the inert gas experiments. Discussions of older recoil experiments with bound parent nuclei can be found in the above mentioned review articles.

The first of the three above mentioned questions which can be illuminated by recoil investigations finds its most clearcut answer in experiments with pure neu-trino emission, i.e. π-decay, μ-capture and electron capture. In such cases essentially a two body decay results showing a line spectrum for the recoil, the line being broadened by thermal motion of the parent substance mainly. The most beautiful demonstration of the missing momentum is found in the π-μ decay discussed in Vol. XLIII.

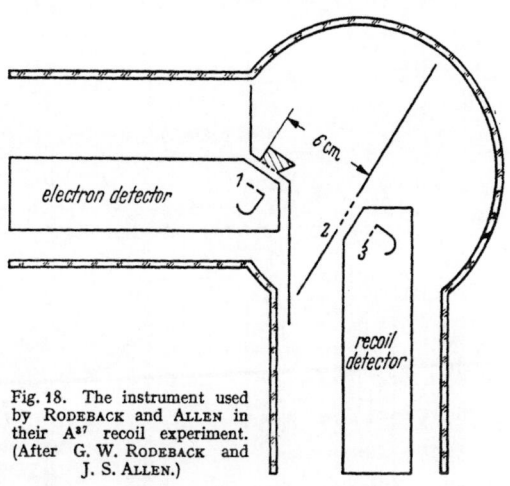

In electron capture experi-ments the most important results have been obtained with A^{37}. This nucleus decays with a half-life of ~ 35 days by electron capture $(K/L = 92/8)$ [3] into

Fig. 18. The instrument used by RODEBACK and ALLEN in their A^{37} recoil experiment. (After G. W. RODEBACK and J. S. ALLEN.)

neutral Cl^{37} recoil atoms. The recoils are left with vacancies in the inner electron shells which are filled through small amounts of X-ray emission and, to a larger extent, by Auger transition. These processes occur inside a very short interval of time and often result in highly charged recoils. Also, the abrupt shaking of the electronic core due to the sudden change of nuclear charge causes some disturbances in the electronic configuration which result in the emission of low energy electrons and the build-up of charge on the recoils. The average charge of the recoils has been measured and found to be 3.2 ± 0.2[1].

The first recoil investigations with this isotope were carried out by RODEBACK and ALLEN[2]. Their instrument is shown in Fig. 18. Delayed coincidences between the fast Auger electrons and the recoil ions are counted and the geometry of the instrument is so arranged that only recoils from that part of the instrument which is cross-hatched in the drawing are accepted when in coincidence with Auger electrons. Thus in this type of instrument the source area is defined by selecting the possible paths of the particles detected. The Auger electrons are recorded on the electron detector, and the recoils are accelerated between the grids near the recoil detector and then detected in the recoil detector. The de-tectors are Allen type multipliers. The distance travelled by the recoil atoms

[1] M. L. PERLMAN and J. A. MISKEL: Phys. Rev. 91, 899 (1953). — S. WEXLER: Phys. Rev. 93, 182 (1954).

[2] G. W. RODEBACK and J. S. ALLEN: Phys. Rev. 86, 446 (1952).

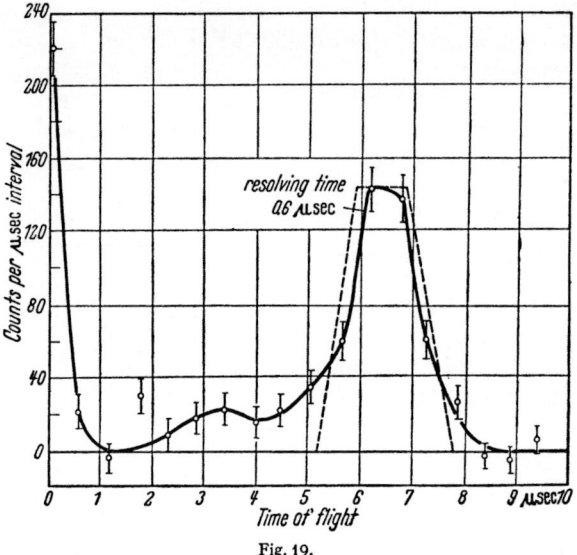

Fig. 19.

Fig. 19. The time-of-flight distribution for the A³⁷ recoils observed by RODEBACK and ALLEN. (After G. W. RODEBACK and J. S. ALLEN.)

Fig. 20. The vacuum chamber and condenser system for the crossed field neutrino recoil spectrometer used for A³⁷ investigations by the author. (After O. KOFOED-HANSEN and A. NIELSEN.)

Fig. 21. Recoil current obtained in the crossed field instrument of Fig. 20 as a function of the magnetic field and for an electric field of 60 volts across the condenser. As the field becomes higher the highly charged ions begin to spiral out between the plates and the current to the collector drops. The separate contributions from the different charges are illustrated. (After O. KOFOED-HANSEN and A. NIELSEN.)

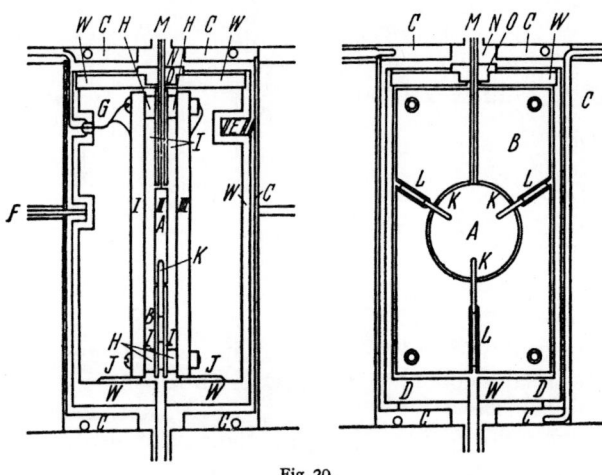

Fig. 20.

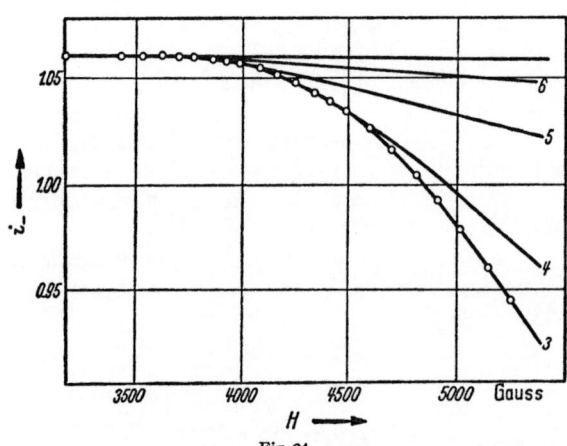

Fig. 21.

and the delay time give the velocity of the recoils. Great care is taken to maintain the shields, baffles etc. at the same electric potential in order not to accelerate or decelerate the low energy recoils and thereby falsify the results. The energy of the recoils is expected to be 9.6 ev from the observed mass difference between A³⁷ and Cl³⁷ of (814 ± 2) kev [1] as measured by nuclear reaction data. The time of flight distribution observed by RODEBACK and ALLEN is shown in Fig. 19. The dotted curve shows the calculated resolution curve for 9.6 ev recoils including the ∼7% velocity spread caused by the thermal motion of the parent nuclei. The agreement is excellent and clearly demonstrates the two-body character of the decay and the momentum taken away by the neutrino.

[1] H. T. RICHARDS, R. V. SMITH and C. P. BROWNE: Phys. Rev. **80**, 524 (1950). — W. A. SCHOENFELD, R. W. DU-BORG, W. M. PRESTON and C. GOODMAN: Phys. Rev. **85**, 873 (1952).

One of us has tried another approach to the problem of recoil measurements. Instead of trying to define a source volume as described in the above

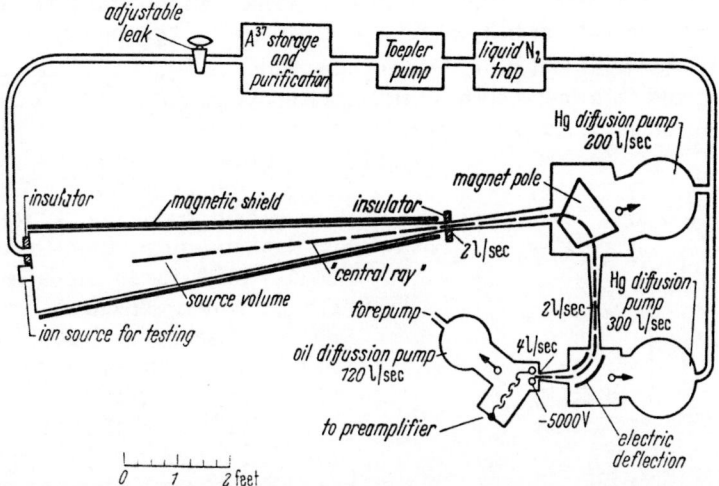

Fig. 22. The recoil spectrometer used by SNELL and PLEASONTON in the A³⁷, Kr⁸⁵ and Xe¹³¹ investigations.
(After A. H. SNELL and F. PLEASONTON.)

experiment or as will be illustrated in the experiments by SNELL and PLEASONTON to be discussed presently, one may let the entire vacuum system be filled with A³⁷ and construct the instrument in such a manner that the geometry may be accounted for in precise manner and without the introduction of simplifying assumptions. The instrument is shown in Fig. 20. The radioactive noble gas is kept between the plates of a condenser and a magnetic field is applied parallel to the plane of the condenser plates. Across the condenser an electric field is maintained and the current of recoils and electrons to the collector, i.e., the central part of one of the electrodes, is measured by means of a vibrating reed electrometer. From this current and its variation with the crossed electric and magnetic fields it is possible to obtain information about the energy and charge of the recoils. The current as a function of the magnetic field in a typical experiment is shown in Fig. 21. The contributions from the various charge values observed are also given. The experimental results[1]

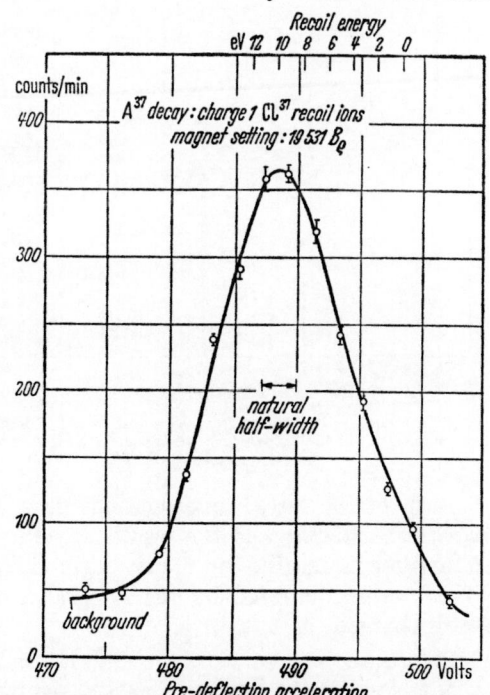

Fig. 23. The peak of singly charged recoils as observed by SNELL and PLEASONTON in the instrument shown in Fig. 22. (After A. H. SNELL and F. PLEASONTON.)

---

[1] O. KOFOED-HANSEN: Phys. Rev. 96, 1045 (1954). — O. KOFOED-HANSEN and A. NIELSEN: Dan. Mat. fys. Medd. 29, No. 15 (1955).

show that the neutrino momentum times c equals (812 ± 8) kev in good agreement with the above quoted nuclear reaction data. Information on the charge distribution was also obtained. The instrument is clearly an integral type of spectrometer which, of course, has several disadvantages if high precision is desired.

A more conventional recoil spectrometer which is capable of extremely good resolution and high precision has been constructed by SNELL and PLEASONTON[1]. Their instrument is shown in Fig. 22. In this instrument the source is defined

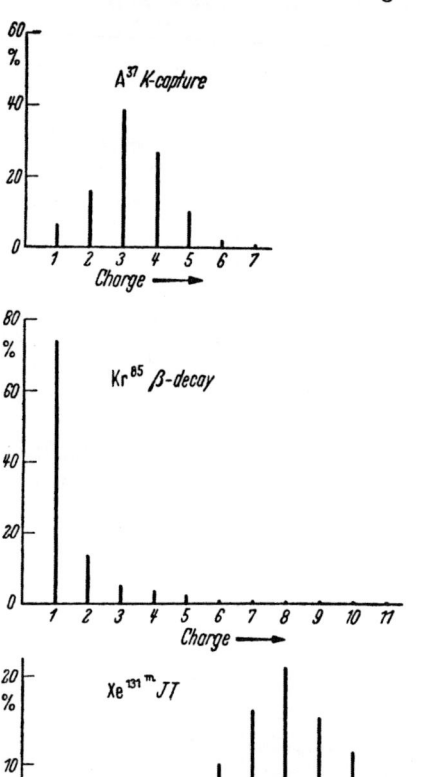

by means of differential pumping. Strong pumping is applied between the narrow exit from the source volume and the narrow exit from the deflection magnet and between the latter and the entrance to the detector. The detector volume is also pumped down with high speed pumps. In order to save the radioactive gas the first mentioned pumps lead into a purification and recirculation system. The energy and charge of the recoil atoms are measured by separate magnetic and electric deflection. The result is a recoil energy of (9.63 ± 0.06) ev in very good agreement with the value expected from the A^{37}-Cl^{37} mass difference. The observed recoil line for the singly charged recoils is shown in Fig. 23 and the charge distributions observed in the decay of A^{37}, Kr^{85} (β-decay) and X^{131} (isomeric transition) are shown in Fig. 24. The main part of the broadening of the line in Fig. 23 is caused by the instrumental resolution and thermal motion. The additional broadening from the emission of the Auger electrons of highest energy was observed on the lines corresponding to charge 2, 3, etc. but not on the singly charged ions.

Fig. 24. The charge distributions from K-capture of A^{37}, β-decay of Kr^{85} and isomeric transition of Xe^{131} observed by SNELL and PLEASONTON with the recoil spectrometer shown in Fig. 22.

Both of the latter investigations gave information about the *charge distribution of the recoils* and the results agree for most of the cases, but the integral method seems to give too many singly charged recoils. This may be due to the difficulties encountered by the integral method in resolving the effects of the lowest charges.

Before we proceed with the description of the experimental results we shall say a few words about the theoretical significance of β-decay recoil investigations for allowed transitions. It is convenient mathematically to describe the angular correlation between the electron and the neutrino rather than between the recoil and the electron. The mathematical transformation to the latter parameters is

[1] A. H. SNELL and F. PLEASONTON: Phys. Rev. **97**, 246 (1955); **100**, 1396 (1955).

entirely trivial but slightly cumbersome[1]. Expressed in terms of the angle ϑ between the directions of the electron momentum p and the neutrino momentum q the correlation has the following form

$$P(\vartheta) \cdot \frac{1}{2} \sin \vartheta \, d\vartheta = \left(1 + \frac{\beta}{W} + \frac{\alpha \, p}{W} \cos \vartheta\right) \cdot \frac{1}{2} \sin \vartheta \, d\vartheta \qquad (6.3)$$

with

$$\beta = 2\gamma \cdot \frac{g_S g_V |\int 1|^2 + g_T g_A |\int \vec{\sigma}|^2}{(g_S^2 + g_V^2) |\int 1|^2 + (g_T^2 + g_A^2) |\int \vec{\sigma}|^2} . \qquad (6.4)$$

It is this term that is called the FIERZ interference term and which gives an effect in the shape of allowed (and also forbidden) spectra. The quantity α is given by

$$\alpha = \frac{(g_V^2 - g_S^2) |\int 1|^2 + \frac{1}{3}(g_T^2 - g_A^2) |\int \vec{\sigma}|^2}{(g_S^2 + g_V^2) |\int 1|^2 + (g_T^2 + g_A^2) |\int \vec{\sigma}|^2} . \qquad (6.5)$$

This magnitude is called the *angular correlation parameter*. The electron-recoil angular correlation and the recoil spectrum are given as a sum of terms of different shape and multiplied by α and β. Thus the recoil investigations give information which depends linearly on these parameters.

The most important results in β-decay recoil investigations have been obtained with the decays of the neutron and of some inert (monatomic) gases. Among the gases only a few isotopes are sufficiently well suited for the conventional sort of recoil experiments. These experiments require a not too low recoil energy in order to facilitate detection and measurement of the energy. Thus only light nuclei with high β-decay energy [cf. Eq. (6.2)] can be used. Furthermore it is necessary that the decay schemes are simple and well known. The isotopes He^6, Ne^{19}, Ne^{23} and A^{35} have been extensively used. A few data on these isotopes are gathered in Table 2.

Table 2. *Characteristics of isotopes used in recoil investigations.*

Isotope	Halflife	E_{max} Mev	Transitions	Type of couplings
$He^6 \xrightarrow{\beta^-} Li^6$	0.82 sec	3.50	$0^+ \to 1^+$	Superallowed G-T
$Ne^{19} \xrightarrow{\beta^+} F^{19}$	18 sec	2.23	$\frac{1}{2}^+ \to \frac{1}{2}^+$	mixed
$Ne^{23} \xrightarrow{\beta^-} Na^{23}$	40 sec	4.39 67% 3.95 32%	$\frac{5}{2}^+ \to \frac{3}{2}^+$ $\frac{5}{2}^+ \to \frac{5}{2}^+$	Superallowed G-T
$A^{35} \xrightarrow{\beta^+} Cl^{35}$	1.8 sec	4.96 93%	$\frac{3}{2}^+ \to \frac{3}{2}^+$	mostly F.

Most β-decays show a mixture of Fermi (scalar and/or vector) interaction and Gamow-Teller (tensor and/or axial vector) interaction but as seen in Table 2 He^6 and Ne^{23} show pure Gamow-Teller interaction (T or A) and A^{35} almost pure Fermi (S or V). This fact accommodates an unambiguous determination of the coupling constants.

The β-decay interaction of Ne^{19} is known to be a mixture of Fermi and Gamow-Teller interactions. From the analysis of ft-values the matrix elements are predicted to give a cancellation of the asymmetry in the β-ν correlation, i.e. $\alpha = 0$ under the assumption that the β-decay interaction is S, T or V, A. Four different experiments using Ne^{19} have been reported giving consistent results. This is a remarkable fact indicating that the methods are good.

ALFORD and HAMILTON[2] used a variable electric potential in order to define their source volume. The potential is triggered by the arrival of the β-particle

[1] O. KOFOED-HANSEN: Dan. Mat. fys. Medd. 28, No. 9 (1954).
[2] W. P. ALFORD and D. R. HAMILTON: Phys. Rev. 95, 1351 (1954).

in the scintillation counter, shown in Fig. 25, and decreases with time so that a recoil particle starting closer to the grid than L_0 will, at the time it arrives at the grid, find a potential too high for it to surmount. On the other hand a particle travelling from the distance L_0 or further away will spend so much time travelling that the voltage will have dropped to such an extent that by the time the recoil particle arrives at the grid it can pass the grid and be registered in the multiplier, i.e., the source volume is defined to the hatched area of Fig. 25. The distribution in transit time measured by delayed coincidences then gives the energy distri-

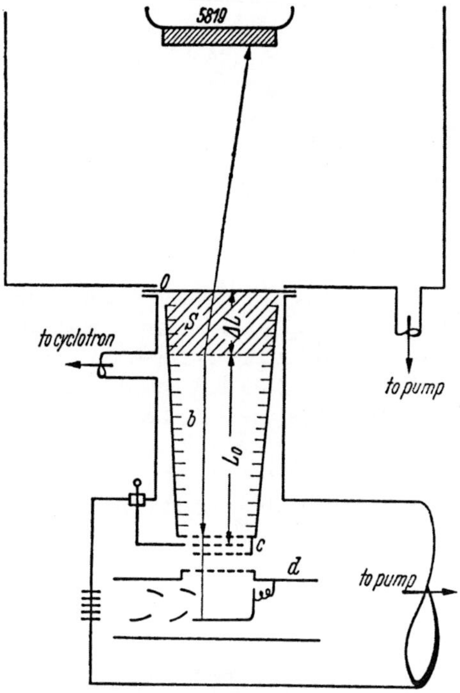

Fig. 25. Recoil spectrometer used by Alford and Hamilton for the study of Ne[19]. (After W. P. Alford and D. R. Hamilton.)

bution for the sensitive volume. In this manner the recoil energy distribution for a fixed angle between recoil and electron (180°) is measured and from this spectrum the value $\alpha = -0.8 \pm 0.4$ was derived.

The second Ne[19] recoil experiment was carried our by Maxson, Allen and Jentschke[1]. Their instrument is shown in Fig. 26. The source volume E is defined by a closed hemisphere facing the 2π plastic scintillation counter, B. The hemisphere is transparent to positrons but stops of course the recoils. The recoils are energy selected in the electrostatic spectrometer G and detected in the system H-J by means of the usual multiplier arrangement. In front of the source volume is placed a gate which can be opened or closed with the handle F. Coincidences are registered with the gate open and closed, and the differences between these measurements gives the desired recoil spectrum. The results gave $\alpha = -0.21 \pm 0.08$.

The instrument used by Good and Lauer[2] for the third Ne[19] experiment is shown in Fig. 27. The source volume is defined by the plastic β-scintillator and the perforated plate in front of the scintillator. The recoil detector is placed opposite to the β-counter. A pump is pumping from the drift chamber (between the perforated plate and the recoil detector) into the source volume. This pump in connection with the pumping resistance of the perforated plate maintains a density of active gas in the source volume 100 times that in the drift chamber. The pulses from the β-counter and the recoil detector are displayed on an oscillograph. The oscillograph is triggered by a coincidence circuit operated from the two detectors. The recoil energy is measured from the time-of-flight as seen on the oscillograph. Only high energy β-events are used (>1.4 Mev). This high energy and the chosen geometry ensures that when a recoil corresponding to a β with $E_\beta > 1.4$ Mev is detected, the corresponding β will also be detected. Background measurements are performed while the movable foil close to the perforated

[1] D. R. Maxson, J. S. Allen and W. K. Jentschke: Phys. Rev. 97, 109 (1955).
[2] M. L. Good and E. J. Lauer: Phys. Rev. 105, 213 (1957).

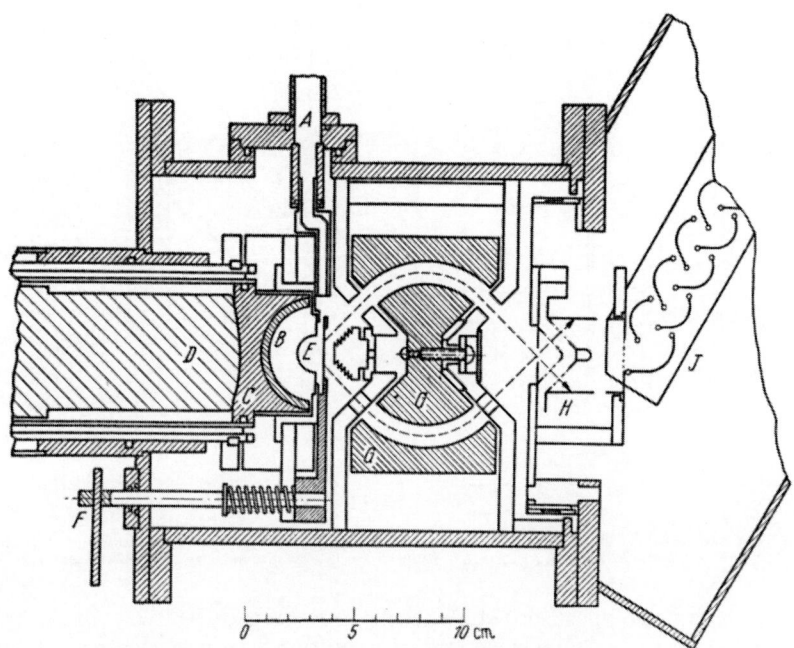

Fig. 26. The recoil spectrometer used by Maxson, Allen and Jentschke in their investigation of Ne¹⁹.
(After D. R. Maxson, J. S. Allen and W. K. Jentschke.)

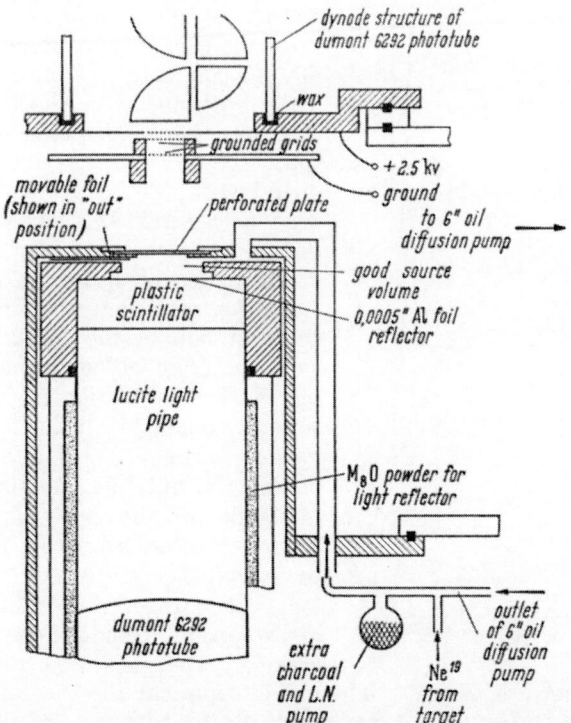

Fig. 27. The instrument used by Good and Lauer in the Ne¹⁹ recoil experiment. (After M. L. Good and E. J. Lauer.)

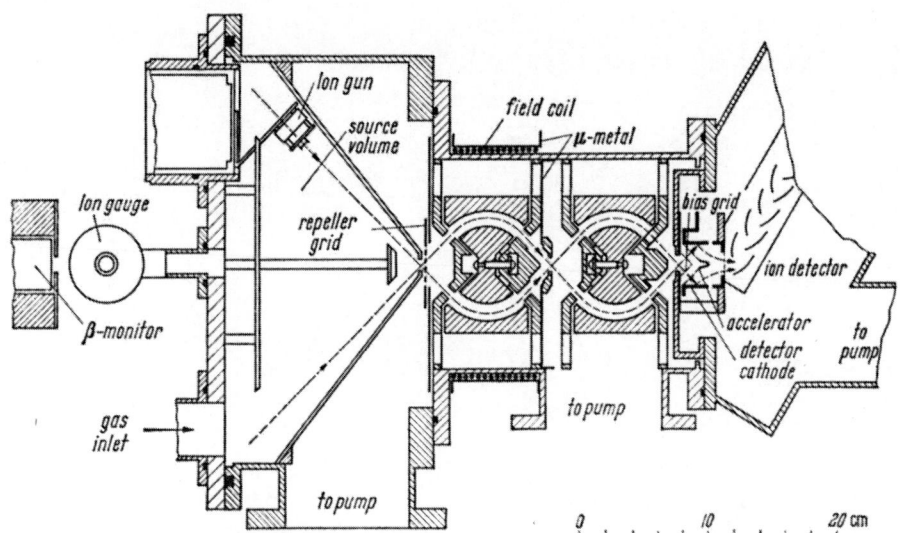

Fig. 28. The second version of the recoil spectrometer constructed by the Allen group and used in the recoil experiments on A^{35}, Ne^{19}, Ne^{23} and He^{6}. (After J.S. ALLEN, R.L. BURMAN, W.B. HERMANNSFELDT, P. STÄHELIN and T.H. BRAID.)

plate (shown in "out" position) stops recoils from the good source volume. As seen the apparatus measures the recoil energy distribution for β-particles of given energy (>1.4 Mev). The measurements are best fitted with the assumption $\alpha = 0.14 \pm 0.13$.

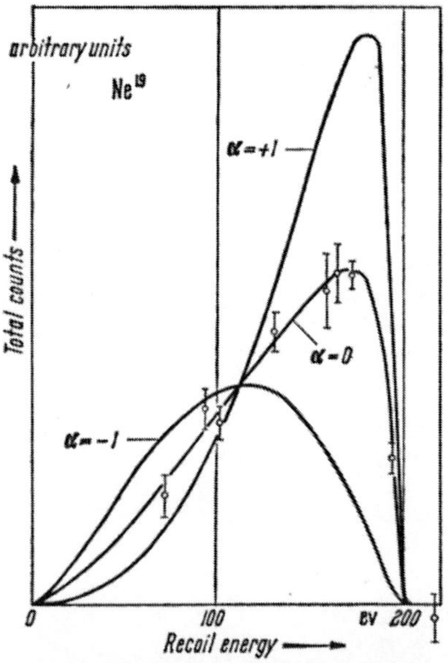

Finally, ALLEN, BURMAN, HERMANNSFELDT, STÄHELIN and BRAID[1] have measured the energy distribution of recoils from any β-ν combination from the decay of Ne^{19}. Their instrument is shown in Fig. 28 and it is seen that only recoils are detected and that the coincidence method has been abandoned. Thus the recoil spectrum as such is measured. The recoil measuring device consists of a double electrostatic spectrometer (cf. also Fig. 26) with a special Allen type recoil detector. The field coil along the first half of the spectrometer prevents electrons from going through the spectrometer while measuring negative ions. The source volume is defined by the conical vacuum chamber walls and the baffle just behind the aperture defining the entrance of recoils into the spectrometer. The background is measured with a potential applied to the repeller grid shown. Thereby recoil ions originating from the source volume are stopped. Strong differential pumping is applied. The measured recoil energy distribution is shown in Fig. 29. The curves represent the theoretical expectation for $\alpha = -1$, 0 and $+1$. The measurements fit with $\alpha = 0.00 \pm 0.08$.

Fig. 29. The results of the Ne^{19} experiment as performed by means of the instrument shown in Fig. 28. The theoretical curves shown are the recoil energy distribution as calculated assuming different values of the angular correlation parameter α. (After J.S. ALLEN, R.L. BURMAN, W.B. HERMANNSFELDT, P. STÄHELIN and T.H. BRAID.)

---

[1] J.S. ALLEN, R.L. BURMAN, W.B. HERMANNSFELDT, P. STÄHELIN and T.H. BRAID: Phys. Rev. **116**, 134 (1959).

This experiment is especially important, because the reasonably good consistence between this experiment and the other Ne^{19} experiments gives us good reason to believe in the other three experiments which have been made with A^{35}, Ne^{23} and He^6 in the apparatus of Fig. 28. These experiments gave the results

$$A^{35}: \quad +0.97 \pm 0.14$$
$$Ne^{23}: \quad -0.37 \pm 0.04$$
$$He^6: \quad -0.39 \pm 0.05.$$

Also PLEASONTON, JOHNSON and SNELL[1] have measured α_{He} with the instrument previously described in connection with the A^{37} experiments (cf. Fig. 22). The result is in agreement with that of ALLEN et al. Their apparatus is able to distinguish ions with different charges. Both singly and doubly charged ions give consistent results. This consistency can not be established by means of the apparatus Fig. 28. Here one has to rely on the upper half of the curve, the low energy part being disturbed by doubly charged ions*.

Earlier experiments on He^6 giving $\alpha = +\frac{1}{3}$ are now believed to be in error[2].

Recently a different approach to the recoil measurements has been tried. The low energy recoil ions being difficult to handle it might be advantageous to use the fact that some *subsequent radiation* emitted rapidly after the β-decay can show effects depending on the recoil momentum. Of course the half life associated with the recoil nucleus must be short compared with the slowing down time of the charged recoil in the source, which may be a solid. An old suggestion is that the decay of Li^8 will show such effects[3]. The recoil nucleus Be^8 breaks up into two α-particles extremely rapidly, presumably within 10^{-21} sec. In the Be^8 frame of reference the two α-particles possess equal momenta in opposite directions. Transformed to the laboratory frame of reference, the α-particle momentum vectors can show a deviation in angle from 180° and they are not necessarily of equal magnitude, both effects due to the recoil momentum. The kinematics of all particles β, ν and two α's can be extracted from a measurement of the momentum vectors of the β-particle and the two α's.

BARNES, FOWLER, GREENSTEIN, C.C. LAURITSEN, T. LAURITSEN and NORDBERG[6] have made a series of measurements with the double purpose: (1) establishing the spin sequence involved in the $Li^8 \rightarrow Be^{8*} \rightarrow 2\alpha$ decay, (2) determining the coupling constants in the β-decay. They measured the angular distribution of the two α's thus establishing the spins $2^+ \rightarrow 2^+$ in the β-decay and excluding S and V coupling. Their results are shown in Figs. 30 and 31 together with theoretical curves for different couplings and spins. The S and V curves may be understood qualitatively. In the vector case, the electron and the neutrino are emitted preferentially in the same direction ($\alpha = +1$) thus giving large recoil momentum and consequently a large deviation from 180°. In the scalar case ($\alpha = -1$) the

---

* *Note added in proof:* The final conclusion from this experiment is $\alpha = -0.33 \pm 0.01$[4]. RIDLEY[5] has recently found $\alpha_{He} = -0.354 \pm 0.04$. He studied the correlation between β-particle- and recoil-energy.

[1] F. PLEASONTON, C.H. JOHNSON and A.H. SNELL: Bull. Amer. Phys. Soc. **4**, 78 (1959) and Oak Ridge Nat. Lab. annual report 1958/59, p. 5, ORNL-2718.

[2] B.M. RUSTAD and S.L. RUBY: Phys. Rev. **89**, 880 (1953); **97**, 991 (1955). — J.S. ALLEN and W.K. JENTSCHKE: Phys. Rev. **89**, 902 (1953).

[3] R.F. CHRISTY, E.R. COHEN, W.A. FOWLER, C.C. LAURITSEN and T. LAURITSEN: Phys. Rev. **72**, 698 (1947).

[4] C.H. JOHNSON, F. PLEASONTON and T.A. CARLSON: Bull. Amer. Phys. Soc. **6**, 227 (1961).

[5] B.W. RIDLEY: Nuclear Phys. **25**, 483 (1961).

[6] T. LAURITSEN, C.A. BARNES, W.A. FOWLER and C.C. LAURITSEN: Phys. Rev. Letters **1**, 326 (1958). — C.A. BARNES, W.A. FOWLER, H.B. GREENSTEIN, C.C. LAURITSEN and M.E. NORDBERG: Phys. Rev. Letters **1**, 328 (1958).

electron and the neutrino are emitted preferentially in opposite directions, thus giving less recoil energy and less deviation from 180°. As shown in Fig. 31 this

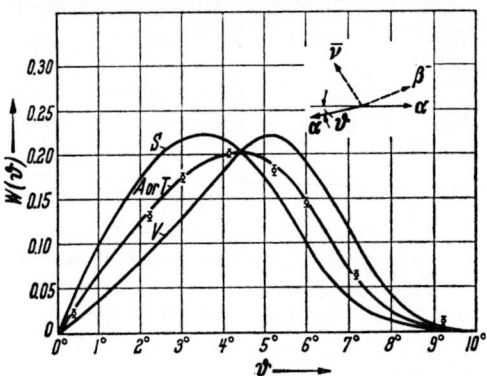

Fig. 30. The angular correlation between the two α-particles emitted during the decay Li⁸→Be⁸*→2α as measured by Lauritsen et al. The curves shown are the theoretical angular distributions calculated for the case of a spin sequence 2⁺→2⁺ and assuming a pure β-coupling being respectively S, V, T or A. (After T. Lauritsen, C.A. Barnes, W.A. Fowler and C.C. Lauritsen.)

experiment is not able to distinguish between A and T because of the spin sequence $2^+ \to 2^+$. To enable this distinction another experiment was performed: A measurement of α-particle momentum distribution in a direction of 180° to the β-particle. The result is that the coupling is essentially axial vector. The possible admixture of T and S-V is claimed to be less than 10% for each.

Also Lauterjung, Schimmer, Schmidt-Rohr and Maier-Leibnitz[1] have used Li⁸ for a similar experiment. They measured the distribution of the difference between the energies of the two α-particles for a definite β-energy interval in a geometry with two α-proportional counters viewing the source separated in angle by 180° and an Anthracene scintillation β-counter at 45° to the line connecting the α-counters. This experiment is sensitive to both type of coupling and spin sequence. Their experiment is in agreement with the Li⁸ experiment described above.

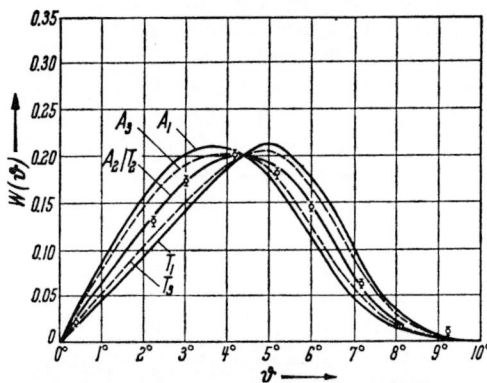

Fig. 31. The experimental points of Fig. 30 compared with theoretical curves corresponding to pure A or T interactions. The parameter (indices) are the spin value assumed for Li⁸. (After T. Lauritsen, C.A. Barnes, W.A. Fowler and C.C. Lauritsen.)

Also a *subsequent γ-ray* can be used. The difficulty in this sort of experiment is that the shift in γ-ray energy is too small to be measured easily. In an experiment made by Burgov and Terekhov[2] this difficulty is overcome by using nuclear resonance fluorescence and using a decay with a cascade of two γ's following the β-decay. The experimental setup is

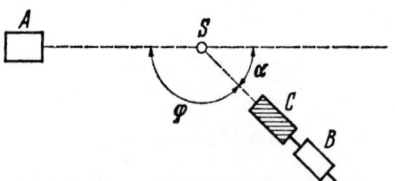

Fig. 32. The geometrical arrangement of detectors and resonance absorber used by Burgov and Terekhov. (After N.A.Burgov and In. V. Terekhov.)

shown schematically in Fig. 32. A and B are γ-detectors and C is the resonance absorbing medium. Coincidences between A and B are measured for different angles. The absorption (the resonance part) will depend on (1) the angle α (2) the recoil momentum after the β-decay and thus the β-decay interaction. The experiment has been performed with Na²⁴. Although the

¹ K.H. Lauterjung, B. Schimmer, V. Schmidt-Rohr and H. Maier-Leibnitz: Z. Physik 155, 547 (1959).
² N.A. Burgov and In.V. Terekhov: J. Exp. Theor. Phys. USSR. 35, 932 (1958). — JETP [translation] 8, 651 (1959).

statistics are rather poor the result is in agreement with axial vector interaction, assuming the interaction to be mostly of the Gamow-Teller type.

The conclusions derived from neutrino recoil experiments are thus that energy is not conserved in β-decay and related processes, and that momentum is missing. Simultaneously, the observed data are in excellent agreement with calculations based on the neutrino hypothesis. It follows quite definitely that β-particle and recoil participate in a direct three-particle disintegration in β-decay and in a two-particle disintegration in K-capture. Finally we may mention the most important conclusion which will be discussed in Chap. E that the G-T interaction is axial vector and that the Fermi interaction is vector.

7. Experiments on direct detection of neutrinos. The experiments described so far have all been connected with the experimental verification of the momentum carried off by neutrinos when leaving a decay act. The neutrino hypothesis was created for this specific purpose however, and from relativistic arguments it is by no means very surprising that momentum and energy disappear simultaneously. In a certain sense one may say that neutrinos are invented in order to save the usual conservation laws and for no other purpose. A firm belief in the real existence of neutrinos can be retained only if neutrinos are observed doing something at a later time than that at which they are created. Possible reactions of this kind may be used for *direct neutrino detection*. By the words "neutrinos doing something" is of course understood "neutrinos giving off energy to observable particles". There are two types of reactions which we are able to imagine as sources of direct neutrino detection and which have a fair amount of physical significance.

Theoretical considerations by HOUTERMANS and THIRRING[1] based on the virtual dissociation of neutrinos into neutron, antiproton and electron suggest that the neutrino possesses a *magnetic moment* of the order of magnitude $f_{th} \sim 10^{-10}$ Bohr magneton. If this is so, neutrinos will undergo elastic collisions with atomic electrons when passing through matter. The cross section for the production of electrons with an energy W (in units $m\,c^2$) in the energy range dW when the incident neutrino has an energy E (in units $m\,c^2$) has been calculated by BETHE[2] and is given by

$$\sigma(W)\, dW = f^2\, \pi\, (e^2/m\, c^2)^2 \left(1 - \frac{W}{E}\right) \frac{dW}{(1+W)\, W} \tag{7.1}$$

for $W \gg B$ where B is the binding energy of the electrons. The earliest experiments designed to observe or set a limit on the magnetic moment f was carried out by NAHMIAS[3]. Using 5 grams of Ra inclosed in 91 cm of lead, NAHMIAS searched for possible ionization from neutrino scattering and was able to set a limit of $f < 2 \cdot 10^{-4}$ BOHR magneton. His experiment was modelled after earlier neutrino absorption experiments carried out by CHADWICK and LEA[4], but he used stronger sources and heavier shielding against γ-rays. Considerable improvements in this limit have since been obtained, but quite a factor is still left before anything like the theoretical value can be measured[5]. BARRETT used methods similar to those of CHADWICK and LEA but with H^3 as the source of neutrinos.

[1] F. G. HOUTERMANS and W. THIRRING: Helv. phys. Acta **27**, 81 (1954).
[2] H. A. BETHE: Proc. Cambridge Phil. Soc. **31**, 108 (1935).
[3] M. E. NAHMIAS: Proc. Cambridge Phil. Soc. **31**, 99 (1935).
[4] J. CHADWICK and D. E. LEA: Proc. Cambridge Phil. Soc. **30**, 59 (1934).
[5] J. H. BARRETT: Phys. Rev. **79**, 907 (1950). — F. G. HOUTERMANS and W. THIRRING: Helv. phys. Acta **27**, 81 (1954) using data from J. L. KULP and L. E. TYRON: Rev. Sci. Instrum. **23**, 296 (1952). — H. R. CRANE: Rev. Mod. Phys. **20**, 278 (1948). — C. L. COWAN jr., F. REINES and F. B. HARRISON: Phys. Rev. **96**, 1294 (1954).

HOUTERMANS and THIRRING and CRANE base their arguments on the flux of neutrinos from the sun derived from the assumption of the nuclear reaction cycles responsible for solar energy. From such considerations it follows that the flux arriving at the earth is $\sim 6 \cdot 10^{10}$ neutrinos per cm² per sec. From geophysical arguments about the heat dissipated in the earth if the cross section had such a magnitude as to be appreciably absorbed in the earth it is possible to deduce that $f < 2 \cdot 10^{-7}$ BOHR magnetons.

COWAN jr., REINES and HARRISON use neutrinos from a nuclear reactor and investigate the number of counts in large liquid scintillators carefully protected against γ-rays and neutrons by heavy shielding. The number of counts with the reactor on and off gives a background counting rate due to the presence of the reactor. An upper limit on the neutrino cross section can be derived from the known neutrino flux and by assuming that the total background is due to neutrinos. From this the limit $f < 10^{-7}$ BOHR magnetons can be derived. The assumption that all counts are due to neutrinos is of course a rough overestimate of the effect from neutrinos and is the reason for the result being an upper limit.

In order to rule out the possibility of neutrinos constituting a health hazard near a reactor, WOLLAN[1] carried out a study of the scattering cross section of neutrinos on hydrogen and found an upper limit of $\sigma < 2 \times 10^{-30}$ cm².

The second reaction which may eventually lead to direct neutrino detection is the *inverse β-decay* which originates from reactions of the type

$$\nu + p \rightarrow n + \beta^+ \tag{7.2}$$

where the proton and neutron may be bound in nuclei. Detailed balancing arguments relate the cross section for such reactions directly to the β-decay probabilities. If the reactions are

$$A^Z \rightarrow A^{Z+1} + \beta^- + \nu \tag{7.3}$$

$$\nu + A^{Z+1} \rightarrow A^Z + \beta^+ \tag{7.4}$$

and if the comparative half-life of the β-decay is ft, we find for the inverse reaction (7.4) the following cross section

$$\sigma = \pi^2 \ln 2 \cdot \frac{\hbar}{m c^2} \left(\frac{\hbar}{m c} \right)^2 \frac{p W}{f t} \tag{7.5}$$

where in the usual inconsistent dimensions used in β-decay the electron momentum p and energy W and the FERMI integral f are measured with $m = c = \hbar = 1$ whereas the remaining quantities including t are given in cgs units. For the inverse neutron decay $(ft = 1200)$ and for a reaction (7.4) resulting in a 1.5 Mev positron the result is $\sigma \sim 9 \cdot 10^{-44}$ cm².

The radioactivity of the product nucleus may be used to detect the occurrence of the inverse β-decay reaction. Large quantities of absorber material purified from the radioactivity in question are left in the neighbourhood of a source of neutrinos for times comparable to the lifetime of the radioactivity. Of course radioactivity will always be produced due to effects from cosmic rays and radiation from the source of types other than that desired. Also the detection of minute amounts of radioactivity and the extraction of such amounts from large quantities of material present several intricate problems. Background counting rates must be minimized and the shielding problems are quite serious. Two different types of approaches to the problem have been used each employing considerable experimental skill and utilizing impressive technical resources.

---

[1] E. O. WOLLAN: Phys. Rev. **72**, 445 (1947).

CRANE[1] used a relatively small neutrino source in his investigation of the reaction

$$Cl^{35} + \nu \rightarrow S^{35} + \beta^-. \tag{7.6}$$

The source consisted of 1 mc mesothorium in equilibrium with its decay products and the target, a three pound bag of NaCl, was left for 90 days (S^{35} half-life 87 days). No positive effect was observed and the sensitivity of the experiment was such that a limit on the cross section of $\sigma < 10^{-30}$ cm² was obtained. Since then the possibilities of the strong neutrino sources from nuclear reactors have become available, and DAVIS jr.[2] has investigated the reaction

$$Cl^{37} + \nu \rightarrow A^{37} + \beta^- \tag{7.7}$$

using 1000 gallon quantities of carbon tetrachloride placed in the vicinity of the Brookhaven reactor. The problem of extracting a few A^{37} atoms from such quantities of liquid in a quantitative manner has been tested with known amounts of A^{37} mixed into the liquid or created directly in the liquid by a (p, n) reaction on Cl^{37} by means of protons produced in the (n, p) reaction on Cl^{35} all in the same liquid. The A^{37} is extracted by sweeping the liquid with He in relatively large quantities and an elaborate system of condensation traps and charcoal traps cooled at different temperatures purifies the He gas and separates A^{37} from carbon tetrachloride vapor, He and impurities of Xe and Kr. The A^{37} is then introduced into very carefully constructed small counters which are shielded against cosmic radiation with an anticoincidence shield of counters. The background counting rate is about 1 count per 4 to 6 minutes and an increase in the counting rate of about 10% of the background counting rate could be detected. The expected cross section is $2 \cdot 10^{-45}$ cm², and it was possible in the experiments of DAVIS jr. to set an upper limit of $\sigma < 2 \cdot 10^{-42}$ cm² to the cross section. Later measurements in the same experiments gave the result $\sigma < 0.9 \cdot 10^{-45}$ cm²/atom[3].

The most succesfull approach to the problem is somewhat less conventional and somewhat more elaborate. REINES and COWAN and CARTER, WAGNER and WYMAN[4] have examined reaction (7.2) on hydrogeneous material using direct coincidence technique.

The process was detected by observing the following events: (1) a β^+ particle is created and brought to rest loosing its initial energy which lies in a certain interval of energy. (2) The β^+ particle annihilates and emits two γ-rays which in turn are stopped and gives photo- or Compton-electrons loosing energy by stopping. For all practical purposes these two events are simultaneous. (3) A neutron is created and is slowed down and diffuses around until it is captured. In the capture the binding energy is emitted as γ-rays. These γ-rays give rise to electrons which may be detected. The neutron capture occurs delayed relative to the initial events. The γ-ray energy is given by the binding energy. This sequence of events is illustrated in Fig. 33 and the instrument used is shown in Fig. 34.

The main difficulty is the rarity of genuine events due to the smallness of the cross section for neutrino absorption. For this reason, heavy shielding against the reactor and cosmic radiation and an anticoincidence shield against penetrating cosmic rays is employed. Also, noise in the electrical equipment is a serious source of trouble and for this reason the following measurements are made:

---

[1] H. R. CRANE: Phys. Rev. **55**, 501 (1939).
[2] R. DAVIS jr.: Phys. Rev. **97**, 766 (1955).
[3] R. DAVIS jr.: Bull. Amer. Phys. Soc., Ser. II **1**, 219 (1956).
[4] F. REINES and C. L. COWAN: Phys. Rev. **113**, 273 (1959). — R. E. CARTER, F. REINES, M. E. WAGNER and J. J. WYMAN: Phys. Rev. **113**, 280 (1959).

(1) The initial pulse of the β^+ and the annihilation radiation is registered, (2) the neutron capture in Cd contained in the liquid scintillator is registered as a delayed coincidence relative to event No. 1, and (3) the total event: pulse from event 1 and time delay and pulse from event 2 is recorded on an oscilloscope and

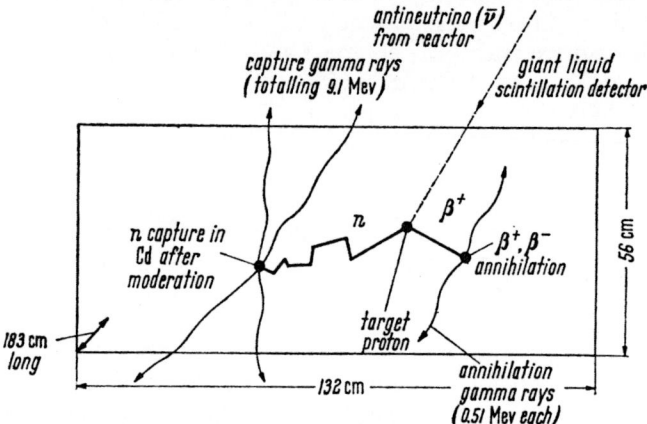

Fig. 33. An illustrative sketch of the sequence of events used by REINES and COWAN in order to identify the process $p + \bar{\nu} \rightarrow n + \beta^+$. (After F. REINES and C. L. COWAN.)

photographed. The true events are found as photographs which show pulses of acceptable magnitude i.e. with energy corresponding to (1) the β^+ plus two annihilation quanta energy range and (2) the neutron capture energy release. From

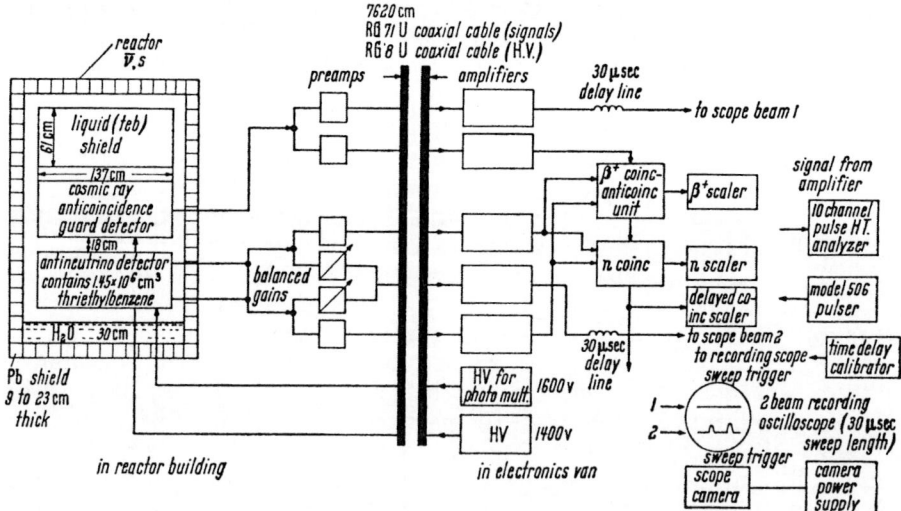

Fig. 34. The experimental arrangement used by REINES and COWAN for direct detection of neutrinos.
(After F. REINES and C. L. COWAN.)

the observed signal rate, the neutrino flux, and the efficiencies for the scintillation counter as regards (1) the β^+ events and (2) the neutron capture, a cross section is obtained.

This cross section is then compared with the calculated cross section. The cross section for monoenergetic neutrinos $\sigma(E_\nu)$ is found from the theory of β-decay and is related to the half life of the neutron [see Eq. (7.5)]. The neutrino spectrum from a nuclear reactor is determined in the following way. The energy distribution of β-rays from fission products is measured. All decays are assumed to show the allowed spectrum shape and the energy distribution is then converted into a distribution of maximum energies. This distribution is then converted into a neutrino energy distribution. It is shown that the choice of nuclear Z to represent the fission products is of minor consequence for the actual result. The resulting cross section is measured to $\sigma = (6.7 \pm 1.5) \times 10^{-43}$ cm²/fission.

Before we compare this result with theory a few words must be said about the scope of the experiments. The most direct result is that the neutrino has been seen directly doing something to a proton, thus proving its material existence. Secondly the cross section is of interest as a test of present day β-theory.

Speculations about the *nature of the neutrinos* regarding their being *Majorana* or *Dirac-particles* were thought to be clarified by cross section measurements. Let us first forget about the polarization of neutrinos in order to understand this argument. If the neutrinos are Majorana particles, i.e. neutrinos and antineutrinos identical, all inverse reactions can be induced by both neutrinos and antineutrinos when special spin phenomena are disregarded. But if neutrinos are Dirac particles one must take care, that the neutrinos used are of the right kind. Otherwise no effect should be expected. Reactor neutrinos are of β^--decay origin. If neutrinos are Dirac particles, these neutrinos should be able to reverse a β^+-decay. This applies to the proton experiment of REINES et al., whereas the reaction (7.7) is not possible. Consequently a negative result of (7.7) with reactor neutrinos as found by DAVIS (p. 33, ref. 3) would reject the possibility that the neutrinos might be Majorana particles. However, with our present knowledge of β-decay, the polarization of the neutrinos also prevent a positive result of the experiment of DAVIS (cf. KONOPINSKI[1]). Thus this experiment should not be able to distinguish between Majorana and Dirac neutrinos, but should simply yield zero cross section in order to be consistent with full neutrino polarization.

The cross section measurements can test the *two component theory* of the neutrino however. As pointed out by LEE and YANG[2], the cross section assuming a two component theory should be twice that assuming a *four component theory*. The result found by REINES et al. is in reasonable agreement with the *two component theory* predicting $\sigma = (6 \pm 1)\ 10^{-43}$ cm²/fission.

Before we leave the subject of direct neutrino detection it should be mentioned that if the neutrino has a magnetic moment the neutrino in β-decay should give rise to bremsstrahlung emission. It has been shown, however, by WENESER[3] that any such effect is far too small to be distinguishable from the ordinary bremsstrahlung due to β-decay. Consequently there is no hope that this kind of investigation could lead to a value for the magnetic moment of the neutrino.

The present trend in speculations on neutrino cross section measurements[4] utilizes the fact that the cross section increases rapidly with increasing energy of the neutrinos. Although difficult such experiments would be very fruitful if they could be carried out by means of present day high energy accelerators.

Concluding the discussion it may be said that the experiments have proved the existence of the neutrino. The cross section is in agreement within experimental error with the two component theory, but the distinction between the Dirac or Majorana neutrinos is more intricate.

[1] E. J. KONOPINSKI: Annual Rev. Nucl. Sci. **9**, 99 (1959).
[2] T. D. LEE and C. N. YANG: Phys. Rev. **105**, 1671 (1957).
[3] J. WENESER: Phys. Rev. **91**, 1025 (1953).
[4] M. SCHWARTZ: Phys. Rev. Letters **4**, 306 (1960). — T. D. LEE and C. N. YANG: Phys. Rev. Letters **4**, 307 (1960). — B. PONTECORVO: J. Exp. Theor. Phys. USSR. **37**, 1751 (1959). — JETP [translation] **10**, 1236 (1960). — F. REINES: Annual Rev. Nucl. Sci. **10**, 1 (1960).

8. Neutrino rest mass. All experimental evidence obtained so far indicate that the *neutrino rest mass* is zero, but of course it cannot be stated that the rest mass is identically zero, only an upper limit has been established.

There are essentially two ways in which values for the neutrino rest mass may be obtained. Firstly, mass determination cycles involving nuclear Q values and β-decay maximum energies can be used because the electron obtains its maximum energy when the neutrino is emitted at rest (or more precisely emitted with the same velocity and in the same direction as the recoil). Thus one finds

$$E_{\beta^-}^{\mathrm{max}} + \mu\, c^2 = (M_{A,\,Z} - M_{A,\,Z+1})\, c^2 \tag{8.1}$$

and the corresponding equation for positron decay, as found by comparison with Eq. (2.10). It should be remembered, however, that the shape of the spectrum depends on the assumption about the mass of the neutrino in such a manner that a word of warning should be said against using Fermi plot extrapolations of the maximum energy for this kind of investigations[1]. If the method of looking at the spectrum directly is used, maximum energies may be inserted in (8.1) only if electrons have actually been observed at the energy in question because in the case of a finite neutrino rest mass the spectrum drops to zero with vertical tangent[2].

A fool-proof example is provided by the decay of H^3 where electrons of 17 kev energy have certainly been observed and where the Q value of the reaction H^3 (p, n) He^3 $[4]$ shows that the mass difference in question is (18 ± 1) kev. Thus it may safely be concluded that the neutrino rest mass is smaller than 2 kev or in units of electron rest mass, that it is smaller than 1/250 of the electron rest mass.

A better method for setting an upper limit on the neutrino rest mass is provided by a close study of the spectrum shape in the vicinity of the maximum energy performed on a decay with as low a maximum energy as possible. Under the assumption of a finite neutrino rest mass the shape of the β-spectrum is given by an expression similar to Eq. (2.4) but with the factor $(W_0 - W)^2$ in that equation replaced by one of the following expressions according to the type of neutrino assumed:

$$\left.\begin{array}{l} \text{Majorana:} \quad (E^{\mathrm{max}} - E + \mu) \qquad\quad (E^{\mathrm{max}} - E)^{\frac{1}{2}}\,(E^{\mathrm{max}} - E + 2\mu)^{\frac{1}{2}}, \\[2mm] \text{Fermi:} \quad \left(E^{\mathrm{max}} - E + \mu\left(1 + \dfrac{1}{W}\right)\right)(E^{\mathrm{max}} - E)^{\frac{1}{2}}\,(E^{\mathrm{max}} - E + 2\mu)^{\frac{1}{2}}, \\[2mm] \text{Dirac:} \quad \left(E^{\mathrm{max}} - E + \mu\left(1 - \dfrac{1}{W}\right)\right)(E^{\mathrm{max}} - E)^{\frac{1}{2}}\,(E^{\mathrm{max}} - E + 2\mu)^{\frac{1}{2}}, \end{array}\right\} \tag{8.2}$$

where μ is the neutrino mass and E^{max} is given by (8.1).

Several careful studies of the upper end of the H^3 spectrum have been performed. Here we shall quote the most recent ones only. Hamilton, Alford and Gross[3] have measured the spectrum in an electrostatic spectrometer constructed by Hamilton and Gross[4] and shown in Fig. 35. The instrument observes the integral spectrum and since a thick target was used the observed spectrum represents the double integral of the spectrum—at least in the vicinity of the maximum energy. For this reason the Fermi plot is here conveniently replaced by the fourth root plot of the electron current observed as a function of the

[1] O. Kofoed-Hansen: Phys. Rev. **71**, 451 (1947). — Phil. Mag. **42**, 1448 (1951).
[2] E. Fermi: Z. Physik **88**, 161 (1934).
[3] D. R. Hamilton, W. P. Alford and L. Gross: Phys. Rev. **92**, 1521 (1953).
[4] D. R. Hamilton and L. Gross: Rev. Sci. Instrum. **21**, 912 (1950).

retarding voltage. The results of HAMILTON, ALFORD and GROSS are shown in Fig. 36 where they are compared with the different curves for different assumptions about the nature of the neutrino and about its mass. For example, curve C corresponds to the Dirac neutrino of rest mass 500 ev. It was concluded from these measurements that the neutrino rest mass is smaller than this magnitude, 500 ev, the Dirac case being the most unfavourable case. In other words the neutrino rest mass is found to be smaller than 1/1000 of the electron rest mass.

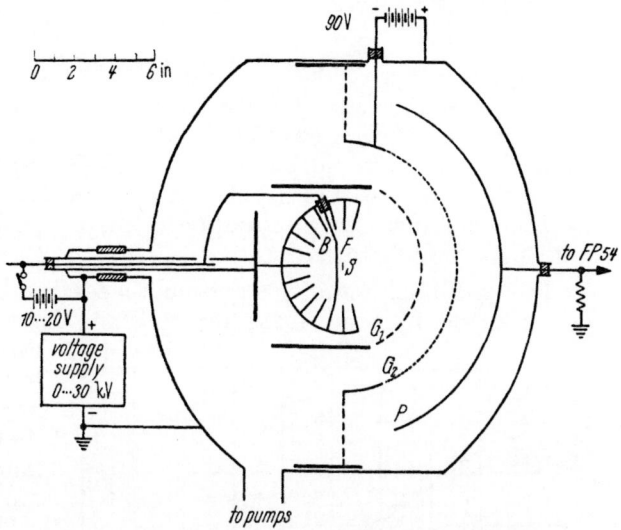

Fig. 35. The electrostatic β-ray spectrometer constructed by HAMILTON and GROSS. (After D. R. HAMILTON and L. GROSS.)

A similar result was found by LANGER and MOFFAT[1] who used a more conventional spectrometer. In both cases, corrections for experimental resolution has been taken care of and a considerable amount of ingenuity has been shown in handling of the low energy electrons from the H^3. Earlier measurements especially by means of proportional counter spectrometers[2], have given similar results. So far, no experiments have given any evidence for a finite neutrino rest mass.

In this connection it can be mentioned that the question whether the *neutrino rest mass is identically zero* has won renewed interest. In the old parity-conserving β-decay theory it was just a bit of information about the fundamental particle, the neutrino. But with the invention of parity-non-conservation it has been

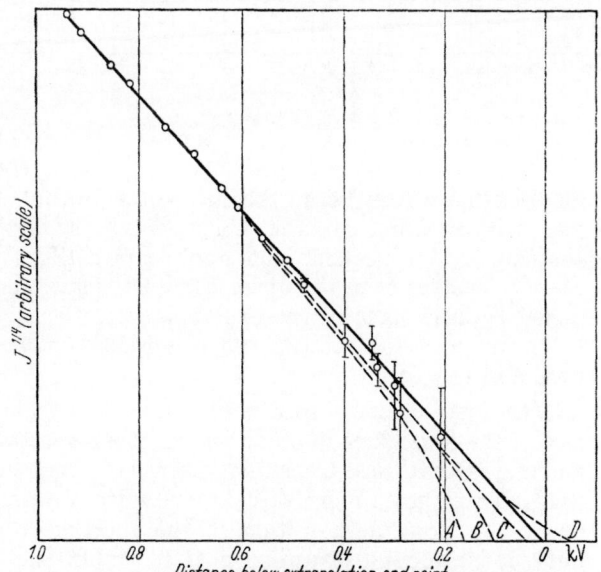

Fig. 36. The fourth root plot of the electron current in the H^3 investigation of HAMILTON, ALFORD and GROSS as obtained with the spectrometer shown in Fig. 35. The measured points are compared with curves for different assumptions about the nature and mass of the neutrino. (After D. R. HAMILTON, W. P. ALFORD and L. GROSS.)

[1] L. M. LANGER and R. J. D. MOFFAT: Phys. Rev. **88**, 689 (1952).
[2] S. C. CURRAN, J. ANGUS and A. L. COCKROFT: Phil. Mag. **40**, 53 (1949). — Phys. Rev. **76**, 853 (1949). — G. C. HANNA and B. PONTECORVO: Phys. Rev. **75**, 983 (1949).

extremely important to establish the identity $m \equiv 0$. The *two component theory*, which is much simpler than the old four component theory and therefore highly attractive, is only valid if $m \equiv 0$.

C. Spectral shapes and ft values.

9. General discussion of shape measurements. In Sect. 3 the measurement of the *maximum energy* of the β-decay was discussed. Furthermore the connection between the *ft-value* and the *order of forbiddenness* was mentioned briefly. In this section the measurements of *spectral forms* shall be discussed in more detail. In this connection also a separation of the $\log_{10} ft$ histogram shown in Fig. 12 will be analyzed further.

The entire subject of β-spectrum shapes has been described in excellent review articles by KONOPINSKI and by WU[1]. There were good reasons that the early measurements of continuous β-spectra were distrusted and their reliability questioned. As a matter of fact the determination of shapes of continuous spectra has been the most difficult problem in β-decay investigations. Considerable effort had to be put into these experiments before reliable spectra and convincing agreement with the Fermi theory were obtained. In fact the spectral shapes in the earlier measurements were so distorted (for experimental reasons which were at that time unknown) that KONOPINSKI and UHLENBECK[2] tried to find a

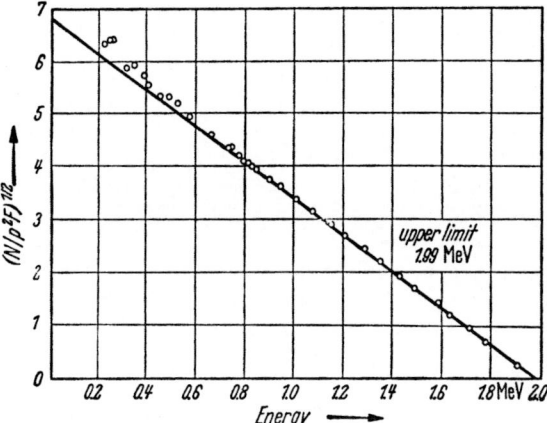

upper limit
1.99 MeV

Fig. 37. The Fermi plot for 72 second In[114] as observed by LAWSON and CORK. (After J. L. LAWSON and J. M. CORK.)

modification of the Fermi theory in order to obtain a more satisfactory agreement with experiments. Their theory was based on the assumption of a derivative coupling for the neutrino and resulted in a $(W_0 - W)^4$ rather than a $(W_0 - W)^2$ energy dependence of the upper part of the spectrum. This theory is now abandoned because some of the experimental difficulties have been overcome with the result that the spectral shape is indeed in very good agreement with the results of the Fermi theory.

The first reliable β-spectrum is usually attributed to LAWSON and CORK[3]. One of the difficulties in obtaining reliable β-spectra for allowed transitions was, and still is, that these transitions are usually very shortlived. LAWSON and CORK were able to get around this difficulty by working with the 50 day In[114] that decays by isomeric transition to the shortlived 72 second In[114] which by an unfavoured allowed transition goes to stable Sn. In this way it was possible for them to obtain a sufficiently longlived source for precise measurements and at the same time work with such a high maximum energy (1,985 Mev) that source troubles are minimized (see Fig. 37). The first good check of the Fermi theory for

---

[1] E. J. KONOPINSKI: Rev. Mod. Phys. **15**, 209 (1943). — C. S. WU: Rev. Mod. Phys **22**, 386 (1950). — Physica, Haag **18**, 989 (1952). — p. 314 of ref [2].
[2] E. J. KONOPINSKI and G. E. UHLENBECK: Phys. Rev. **48**, 7 (1935).
[3] J. L. LAWSON and J. M. CORK: Phys. Rev. **57**, 982 (1940).

low energy spectra was obtained by ALBERT and WU[1]. They investigated the effect of the source thickness and were able to produce extremely thin sources of S^{35} which has a maximum energy of only 0.1670 Mev. With a 1 μg/cm² source the Fermi plot was straight down to the window cut-off at about 20 kev. The results of ALBERT and WU are shown in Fig. 38 where the effect of source thickness is immediately evident.

Meanwhile these are not the only difficulties. Deviations from the linear Fermi plot may occur for several experimental reasons.

α) If the defining slits of the entrance and the excit diaphragms of the β-ray spectrometer are partially transparent to the β-rays (especially near the edge of the slits), particles of high energy may enter the counter to a larger extent than those of low energy. On the other hand high energy particles may be slowed down in the entrance slit and thereby be focused on the counter at a lower energy, giving an excess of low energy particles. This effect can cause either an upward or a downward curvature of the Fermi plot. Good methods for eliminating these effects are difficult to prescribe in general apart from the application of several slit sizes and thicknesses and careful design of the surface of the slit.

β) The finite thickness of the counter window will cut off the low energy portion of the spectrum and produce a downward deviation at low energies from the linear Fermi plot. An example is found in Fig. 38. Measurements of window absorption

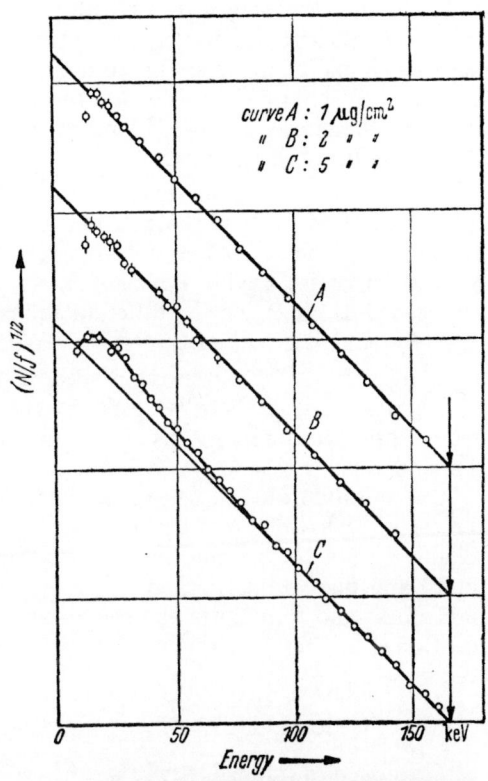

Fig. 38. Fermi plots of the β-ray spectrum of S³ obtained by ALBERT and WU for various thicknesses of the source. (After R. D. ALBERT and C. S. WU.)

may be performed by varying the window thickness and performing an extrapolation to zero window thickness. However, this procedure cannot yield results which are reliable for energies below those for which a considerable effect is caused by the thinnest window applied. Measurements of window absorption may also be performed with artificially accelerated electrons of known intensity. A third method is occasionally applied where a spectrum is measured for which the shape is believed to be accurately known theoretically down to low energies. The correction is found as the ratio between measured points and the theoretical curve extrapolated downwards from those parts of the spectrum where the window correction is known to be negligible. This method is somewhat dubious and should at least never be used where a test of the theory is involved in the measurements. Post-acceleration of the electrons have been suggested as a means of overcoming the difficulties, but unfortunately secondary electron emission causes

[1] R. D. ALBERT and C. S. WU: Phys. Rev. **74**, 847 (1948).

troubles. Pre-acceleration results in a very undesirable complication of the evaluation of the variation of resolution with energy.

γ) Scattering of electrons in source, source backing, baffles, slits and spectrometer walls may cause such electrons to loose energy or direction and consequently to be registered at wrong parts of the spectrum. The results of such effects will generally be to raise the Fermi plot above the straight line. This effect is illustrated by the investigation of Albert and Wu shown in Fig. 38. Several detailed investigations of such effects have been performed[1] and there is a certain inconsistency in the recommendations of the number of baffles which are most efficient. However, one point is certain: the source and the source backing should be made as thin and uniform as possible. Precipitated or dried-in sources may contain clusters of 100 times the average thickness due to crystallization and therefore evaporated sources or sources prepared in isotope separators are generally to be preferred.

δ) Secondary electrons may be ejected when the β-particles which are not focused reach the walls of the instrument or are caught in the baffles or diaphragms. If flat spectrometers or lens spectrometers with baffles designed to differentiate between particles having opposite directions of rotation around the lens axis are used, this effect will be different for positron and negatron emission due to the different directions of the electron paths.

The technique of measuring β-spectra has been greatly improved during the last decade. Today there is no doubt that the β-spectra can be described in a good approximation by means of the formulas Eq. (2.4) to (2.8) using the relevant shape factor C_n.

One question always has to be answered when comparing a spectrum with the theory: what does the shape factor C_n look like in the actual case. The most reliable answer to this question requires knowledge of the spin and parity of the initial and final states. When these characteristics are known the order of forbiddenness and sometimes the shape factor is known. The selection rules were mentioned in Sect. 2, Table 1:

allowed: $\Delta I = 0, 1$; no parity shift

1st forbidden: $\Delta I = 0, 1, 2$; parity shift

n-th forbidden: $\Delta I = n, n+1$; $\begin{cases} \text{parity shift for } n \text{ odd} \\ \text{no parity shift for } n \text{ even.} \end{cases}$

In principle the shape factor C_n can be determined when the selection rule working for the particular transition is known. In practice, however, the theoretical determination of the shape factor is not quite so simple. Usually different terms of different energy dependence are coupled together by means of some nuclear matric elements which are not well known. But in some cases: allowed decays and unique n-th forbidden transitions, that is n-th forbidden transitions with $\Delta I = n+1$, only a single term contributes and the shape factor is known.

The ft value gives a clue to the order of forbiddenness in cases where spin and parity changes are unknown. In Sect. 3, Fig. 12 gives a histogram of $\log_{10} ft$ values for some β-transitions, the spin sequences of which are known. It is seen that the classification according to ft value is not very safe. E.g. the allowed and the first forbidden groups strongly overlap.

In Fig. 39 the histogram is separated in different components. Fig. 39a is the group of superallowed transitions, b and c are the mirror transitions and the

---

[1] e.g. G. E. Owen and C. S. Cook: Rev. Sci. Instrum. 20, 768 (1949).

$0 \rightarrow 0$ (no parity shift) transitions (both superallowed) respectively. Fig. 39d is the allowed unfavoured transitions. The theoretical significance of the distinc-

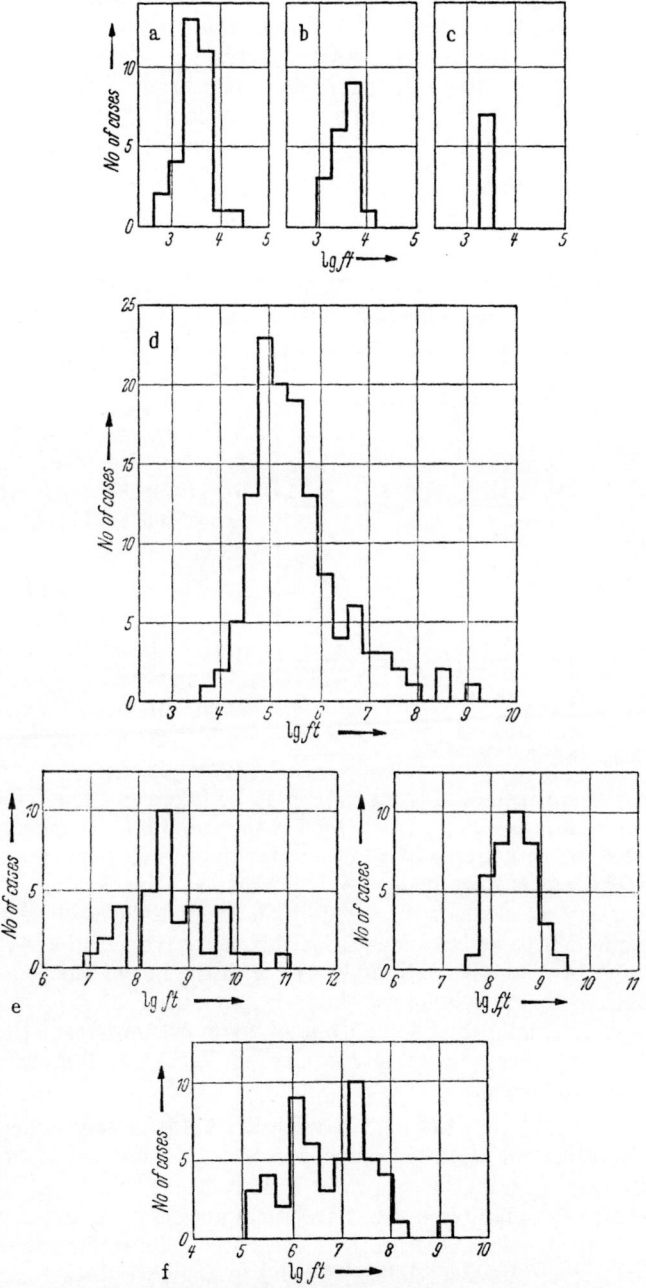

Fig. 39. a $\log_{10} ft$ histograms for superallowed transitions; b for mirror transitions; c for $0 \rightarrow 0$ transitions; d for allowed unfavoured transitions; e $\log_{10} ft$ and $\log_{10} f_1 t$ histograms for unique shape transition; f $\log_{10} ft$ histogram for non-unique shape, first forbidden transitions.

tion between *superallowed* and *allowed unfavoured* transitions is connected with the strictness of certain selection rules like isotopic spin selection rules. More precisely it may be stated that a superallowed transition is a transition between states belonging to the same partition[1] whereas the states connected through an allowed unfavoured transition belong to different partitions. Fig. 39e shows the $\log_{10} f_0 t$ distribution of the unique first forbidden transitions and the $\log_{10} f_1 t$ distribution of the same group. It is interesting to notice that the spread in $\log_{10} ft$ value is much less when the correct spectral shape is used as in the $\log_{10} f_1 t$. Fig. 39f shows the non-unique first forbidden transitions. The second forbidden and the third forbidden transitions are well separated already in Fig. 12. In the next sections the spectral shapes will be described further.

10. Allowed transitions. It is of interest to describe some more examples of spectral forms in order to see how well the theory works for transitions of strongly differing ft values.

Figs. 37 and 38 show examples of allowed unfavoured transitions, namely $In^{114} (E_{Max} = 1.99 \text{ Mev}, \log_{10} ft = 4.5)$ and S^{35} $(E_{Max} = 0.168 \text{ Mev}, \log_{10} ft = 5.0)$.

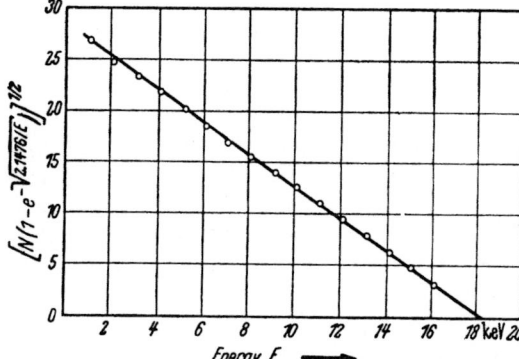

Fig. 72 (Sect. 17α) shows the neutron decay β-spectrum (super-allowed, $E_{Max} = 0.782 \text{ Mev}, \log_{10} ft = 3.7$). Fig. 40 shows the H^3 β-spectrum (superallowed, $E_{Max} = 18.5 \text{ kev}, \log_{10} ft = 3.5$)[2]. All these spectra show reasonably straight Fermi-plots.

The most significant result to be obtained from the $\log_{10} ft$ histograms shown in Fig. 39 of Sect. 9 is the very large spread in ft-values even for transitions of the same order of forbiddenness. The deviation of $\log_{10} ft$ from the mean

Fig. 40. The Fermi plot of H^3 as observed by Curran, Angus and Cockroft. (After S. C. Curran, J. Angus and A. L. Cockroft.)

value 3.5 of the superallowed transitions may be taken as a rough measure of the magnitude of the *unfavoured factor* which may amount to as much as 10^5.

The most striking example of a large unfavoured factor is represented by the decay of C^{14}, the $\log_{10} ft$ value being 9.0. It is believed at present that the smallness of the nuclear matrix element is caused by an accidental cancellation of different matrix element components arising from different terms in the wave function[3]. The β-spectrum of C^{14} agrees extremely well with the Fermi theory for an allowed decay and both of the spins involved have been measured. The important technical applications of C^{14} make this isotope one of the most important tools for tracer investigations and there exists a considerable technical literature about the handling of this radioactivity.

It should be noted that the high ft value for C^{14} is closely followed by many other high ft values for allowed transitions. Another important example is P^{32} with $\log_{10} ft = 7.9$. Again the spectrum shape checks with the Fermi theory. This isotope also has many important technical applications and is readily available in nearly all physical laboratories. Because of its convenient half-life and maximum energy and its availability, P^{32} and to a slightly lesser extent C^{14} serve as standard sources which are nearly always used in checking β-ray spectrometers when continuous spectrum work is intended. In this connection a special

[1] J.M. Blatt and V.F. Weisskopf: Theoretical nuclear physics, p. 718ff. New York: Wiley 1952.
[2] S.C. Curran, I. Angus and A.L. Cockroft: Phil. Mag. **40**, 53 (1949).
[3] R. Sherr, J. B. Gerhart, H. Horie and W. F. Hornyak: Phys. Rev. **100**, 945 (1955).

point in connection with reactor produced P³² is important. P³² is usually produced in a S³² (n, p) P³² reaction. The Q value is so low (~ -0.9 Mev) that
P³² is produced by the more energetic neutrons in the reactor. For many years
it was difficult to obtain linear Fermi plots from reactor-produced P³². It turned
out that the difficulties were caused by the presence in the sample of P³³ produced by the S³³ (n, p) P³³ reaction[1] which has a positive Q value. The half-life
of P³² is 14.3 days and that of P³³ is 25 days. Consequently, the ratio of P³³ to

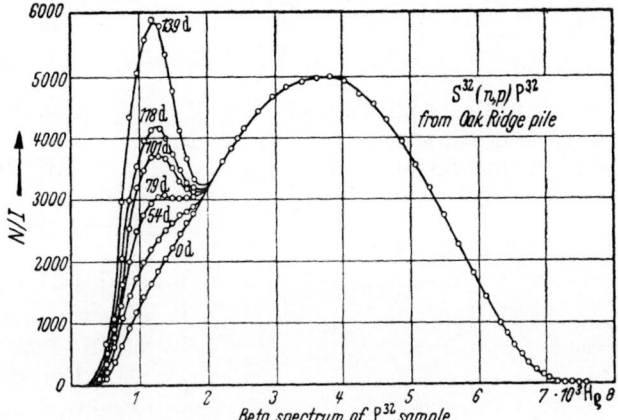

Fig. 41. The growth of P³³ impurity in a pile produced P³² source is shown by means of P³² + P³³ spectrum taken at
different times. (After E. N. JENSEN, R. T. NICHOLS, J. CLEMENT and A. POHM.)

P³² in a reactor-produced sample depends on time and the deviation of the
spectrum from a simple allowed shape could be shown to vary in time as shown
in Fig. 41. This point is of course quite important when P³² is used as a laboratory standard.

Another interesting result of the $\log_{10} ft$ distribution of Fig. 39 is the existence
of several "unfavoured" transitions with very low $\log_{10} ft$ values. These values
are hard to distinguish from some of the
values attributed to superallowed decays and
they are classified as unfavoured transitions
merely because they belong to decays of
relatively heavy nuclei. The cases of
$\log_{10} ft < 4.4$ are summarized in Table 3. No
immediately apparent feature is common to
these transitions. It may be noted that five
cases represent $0^+ \rightleftharpoons 1^+$ transitions, but the
use of this as a rule for finding fast transitions
is ruled out by the glaring exceptions to such
a rule provided by the decays of C¹⁴ and P³².

Table 3. *Fast unfavoured transitions
of $\log_{10} ft < 4.4$.*

Transition	$\log_{10} ft$	Spin	Decay
Nd¹⁴⁰	3.7	$0^+ \rightarrow 1^+$	K
Pr¹⁴⁰	4.3	$1^+ \rightarrow 0^+$	β^+
In¹¹²	4.1	$1^+ \rightarrow 0^+$	β^-
Ru¹⁰⁶	4.3	$0^+ \rightarrow 1^+$	β^-
Mo¹⁰²	4.2	$0^+ \rightarrow 1^+$	β^-
Mo⁹¹	4.0	$9/2^+ \rightarrow 9/2^+$	β^+
Ge⁷¹	4.3	$1/2^- \rightarrow 3/2^+$	K
Zn⁶⁹	4.4	$1/2^- \rightarrow 3/2^-$	β^-

Concluding this section we may remark that the allowed unfavoured transitions show a remarkable spread in $\log_{10} ft$ values and that the highest and lowest
values are based on extremely reliable experimental information. Thus, although
some of the classifications included in Fig. 39 may be in doubt, this main result
is firmly established. The spectral shapes agree well with the theory, and the
lack of a good interpretation of the ft values must be blamed on the lack of a
satisfactory nuclear theory rather than on faults in the theory of β-decay.

[1] R. K. SHELINE, R. B. HOLTZMAN and C. Y. FAN: Phys. Rev. **83**, 919 (1951). — E. N.
JENSEN, R. T. NICHOLS, J. CLEMENT and A. POHM: Phys. Rev. **85**, 112 (1952).

11. Forbidden transitions. *α) Unique shape transitions.* As mentioned in Sect. 3 the unique shape transitions are those forbidden transitions which as a group are most easily recognized experimentally. The occurrence of unique shape transitions furnishes a very good check on the entire theory of β-decay, not only as regards general ideas but also specifically as regards the theory of forbidden decay. When Coulomb effects are neglected, the energy dependence of the correction factors C_n of Eq. (2.4) is given by

$$C_1 = (W^2 - 1) + (W_0 - W)^2, \tag{11.1}$$

$$C_2 = (W^2 - 1)^2 + (W_0 - W)^4 + \tfrac{10}{3}(W^2 - 1)(W_0 - W)^2. \tag{11.2}$$

These factors give rise to quite large changes of the spectral shapes relative to the spectra of allowed transitions.

In 1942 Haggstrom[1] found the unique shape spectra of Rb^{86} and Sr^{89}, but the fact was not recognized because she described the spectra as having "nearly the allowed shape". The first recognition of a unique shape spectrum was made by Langer and Price[2] who measured the spectrum of Y^{91}. The spectrum of Y^{91} is shown in Fig. 42 as a conventional Fermi plot (upper curve) and also as plotted with C_n given by (11.1) (lower curve). The agreement with theory is excellent. At the present time more than 40 cases of unique shape transitions are known. For these transitions one may apply Eq. (2.17) in order to obtain $\log_{10} f_1 t$ values.

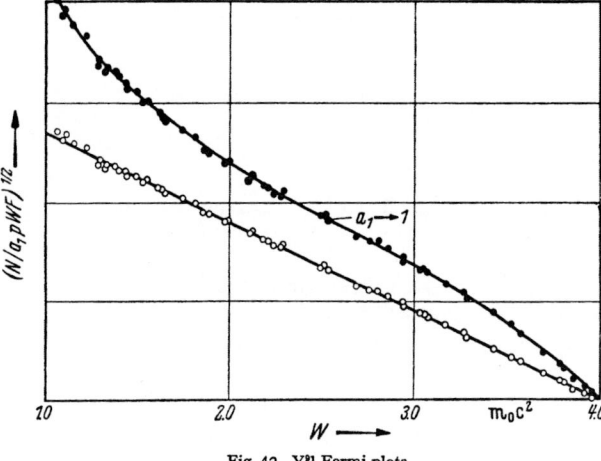

Fig. 42. Y^{91} Fermi plots.

This was first done by Davidson jr.[3] who found that $\log_{10} f_1 t$ values show a much smaller spread than the corresponding $\log_{10} f t$ values for these transitions. This is again a good verification of the theory since it shows that the energy dependence also gives a reasonable effect in the $f t$ value. The effect is illustrated in Fig. 39e. This figure clearly shows the above mentioned property. It is seen that the spread is reduced by a factor ∼100 when the correct energy dependence is taken into account in the Fermi integral.

Some unique shape, second forbidden decays have been found. An example is Be^{10}, which is a $0^+ \rightarrow 3^+$ transition. The Be^{10} spectrum has been measured and agrees very well with theory[4]. The $\log_{10} f t$ value is 13.6. Other examples are Na^{22} $(3^+ \rightarrow 0^+; \log_{10} f t \sim 13)$ and Co^{60} $(5^+ \rightarrow 2^+; \log_{10} f t \sim 13)$.

One unique shape, third forbidden transition, namely the $4^- \rightarrow 0^+$ transition of K^{40} has been established. The spin 4 has been measured and the spin 0 inferred from the fact that the state is the ground state of an even-even nucleus. The $\log_{10} f t$ value is 18.1. Quite thick sources have to be used in order to observe

[1] E. Haggstrom: Phys. Rev. **62**, 144 (1942).
[2] L. M. Langer and H. C. Price jr.: Phys. Rev. **76**, 641 (1949).
[3] J. P. Davidson jr.: Phys. Rev. **82**, 48 (1951).
[4] L. Feldman and C. S. Wu: Phys. Rev. **76**, 697 (1949). — C. S. Wu and L. Feldman: Phys. Rev. **76**, 698 (1949); **87**, 1091 (1952).

the β-ray spectrum since the half-life is 1.3×10^9 years. A considerable amount of experimental ingenuity is displayed in the many publications concerning this isotope (for references see [3]). Since self-absorption effects in the source have to be considered the spectrum is rather uncertain. However, as far as the experimental accuracy goes, the spectrum agrees well with theoretical expectations.

β) *General remarks on non-unique shape transitions.* In general the first forbidden spectra of non-unique or mixed type, i.e., of $\Delta J < 2$, have a form strikingly similar to the allowed shape. An illustrative example is provided by the decay of Pm^{147}. The conventional Fermi plot is shown in Fig. 43. The spin and parity change is expected to be $^5/_2{}^+ \rightarrow ^7/_2{}^-$ from shell model considerations and from the measured spin of $^7/_2$. That these spectral shapes are the allowed shape is in very good agreement with the theory. However, the rule involved is not strict if the nuclear matrix elements accidentally get such signs and values that the major terms in C_n largely cancel. In such cases small differences between the major terms, smaller terms, finite size corrections, finite wavelength corrections and screening corrections will give rise to marked deviations from the allowed shape. The famous example of such a decay is that of RaE which is described in detail in Sect. 11γ.

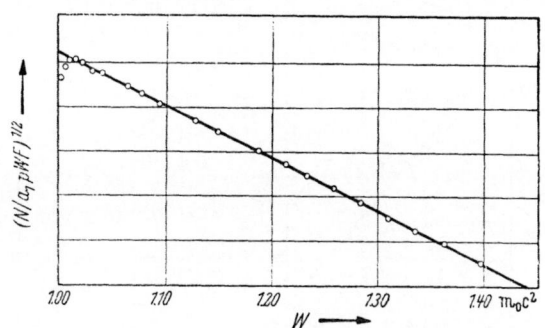

Fig. 43. Conventional Fermi plot for the first forbidden spectrum o Pm^{147}. (After E. J. Konopinski and L. M. Langer.)

An explanation of its shape has been found only on the grounds of such an accidental cancellation.

For second and third forbidden, non-unique shape spectra theoretical deviations from the allowed spectrum form are expected and are found experimentally. However, it should be mentioned that the more highly forbidden the spectra one is dealing with, the more difficult are the problems of avoiding self-absorption and scattering in the source material.

The $\log_{10} ft$ distribution for non-unique shape, first forbidden transitions is shown in Fig. 39g. The most striking feature of this curve is its similarity with Fig. 39d for allowed unfavoured transitions. A quite surprising group of $\log_{10} ft$ values below 6 appears. These transitions play the role for forbidden transitions that the group of unfavoured allowed transitions with $\log_{10} ft < 4.4$ listed in Table 3 play for allowed transitions. They all occur in heavy elements, and their nature as first forbidden decays is quite well established. For one of them, Au^{199}, the spin of the parent nucleus has been measured[1].

Just as for the allowed transitions of low ft values no easily recognizable systematics is apparent for these transitions.

The fact, that $\Delta J = 0$ or ± 1 transitions of either parity show very much the same $\log_{10} ft$ histogram except for the superallowed transitions and the same spectral distribution, means that such ft values do not in general permit the determination of the parity change involved in these transitions.

Among the twice forbidden spectra a group of transitions with $\log_{10} ft \sim 12$ appears. One of the most carefully studied examples is the decay of Cl^{36} which

---

[1] J. B. Reynolds, R. L. Christensen, D. R. Hamilton, A. Lemonick, F. M. Pipkin and H. H. Stroke: Phys. Rev. **99**, 613 (1955).

shows a $2^+ \rightarrow 0^+$ transition of $\log_{10} ft = 13.5$. The shape of the spectrum has been examined by Feldman and Wu[1]. Their result is shown in Fig. 44 in the form of a conventional Fermi plot, and a corrected Fermi plot, the latter using the proper C_n obtained from a ratio of $|A_{ij}/T_{ij}|^2 = 18$, is a straight line.

The $\log_{10} ft$ value 8.6 for the assumed $^1/_2{}^+ \rightarrow {}^5/_2{}^+$ transition in Mg²⁷ is an example of a very low ft value for second forbidden transitions.

A few highly forbidden transitions are known. For example Rb⁸⁷ decays by a non-unique shape, third forbidden transition of spin change $^3/_2{}^- \rightarrow {}^9/_2{}^+$ with a $\log_{10} ft$ of 17.6, and In¹¹⁵ shows a non-unique shape, fourth forbidden transition with a spin change $^9/_2{}^+ \rightarrow {}^1/_2{}^+$ and a $\log_{10} ft \sim 23$.

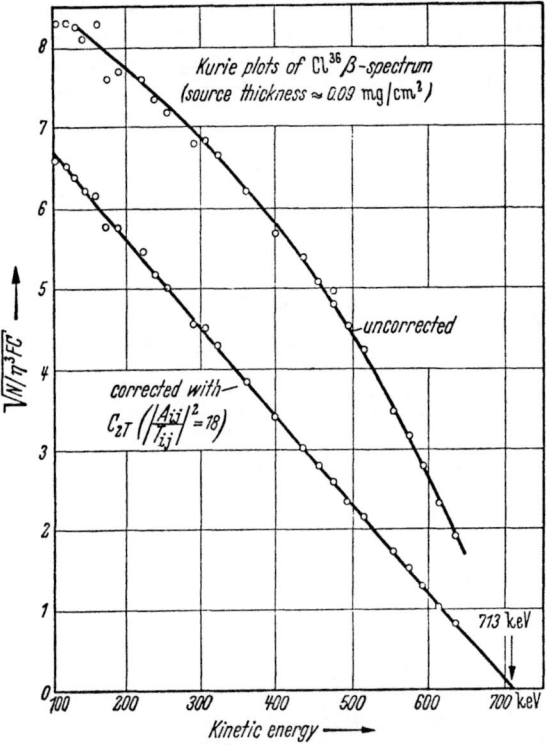

γ) *The decay of RaE.* The decay of RaE has played an extremely important role in the history of β-decay. The decay is simple and no γ-rays from levels in Po²¹⁰ are observed. The decay was therefore used in the calorimetric investigations of mean energies described in Sect. 5 and also in the β-ray charge determinations described in Sect. 1. In spite of the fact that the RaE decay was one of the first recognized nuclear β-decays and the first with a simple decay the interpretation of the spectral shape has been an extremely difficult problem.

Konopinski[2] found that the shape could be nicely fitted by a second forbidden decay with tensor interaction and with the assumption of a spin change $2 \rightarrow 0$ (no). His work was based on the

Fig. 44. Conventional and corrected Fermi plots for Cl³⁶.
(After C. S. Wu, Kai Siegbahn.)

spectra obtained by Flammersfeld, Neary, and Langer and Whitaker[3]. However, the shell model indicates quite definitely, that a parity change is involved in the decay. Consequently, the decay must be considered first forbidden rather than second forbidden. The fact that the $\log_{10} ft$ value is 8.0 would also indicate a first forbidden rather than a second forbidden transition, although with a certain amount of ambiguity (cf. Sect. 11β).

Petschek and Marshak[4] then stated that it was impossible to fit the spectrum unless one assumed a decay of the type $0 \rightarrow 0$ (yes) and the presence of quite a large amount of pseudoscalar interaction in the β-decay coupling. Their considerations were based on more recent spectrum investigations and on the necessary assumption of an accidental cancellation of all major terms in the proper C_n (see

¹ L. Feldman and C. S. Wu: Phys. Rev. 87, 1091 (1952).
² E. J. Konopinski: Rev. Mod. Phys. 15, 209 (1943).
³ A. Flammersfeld: Z. Physik 112, 727 (1939). — G. J. Neary: Proc. Roy. Soc. Lond., Ser. A 175, 71 (1940). — L. M. Langer and M. D. Whitaker: Phys. Rev. 51, 713 (1937).
⁴ A. G. Petschek and R. E. Marshak: Phys. Rev. 85, 698 (1952).

Sect. 11 β). All the many correction terms had to be taken into account, also. The result was questioned because the magnitude of the pseudoscalar coupling constant needed was rather large and because the C_n calculations performed previously were in doubt in the special case of the pseudoscalar coupling[1]. The pseudoscalar coupling constant would, if very large, bring in changes in allowed spectral shapes and from the experimental shapes of He^6 and B^{12} it was possible to show only that $g_P < 50 g_T$ (cf. Sect. 12). The result of PETSCHEK and MARSHAK was finally completely ruled out because the RaE spin was found to be 1.

PLASSMANN and LANGER, and WU and coworkers[2] then showed that a fit may be obtained to a $1 \to 0$ (yes) transition by means of the $S - T$ interaction if the following matrix element ratios are used

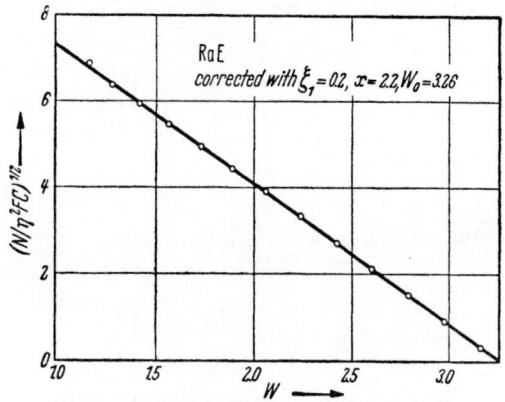

$$i \frac{g_S}{g_T} \frac{\int \beta \vec{r}}{\int \beta \vec{\sigma} \times \vec{r}} = 0.22, \qquad (11.3)$$

$$\frac{2R}{\alpha Z} \frac{\int \beta \vec{\alpha}}{\int \beta \vec{\sigma} \times \vec{r}} = 1.51. \qquad (11.4)$$

These values lead to an accidental cancellation of the large energy independent terms in C_n usually encountered in non-unique shape spectra. The Fermi plot obtained is shown in Fig. 45. It is seen that an excellent fit is obtained.

Fig. 45. RaE Fermi plot on the assumptions (11.3), (11.4). (After E. A. PLASSMANN and L. M. LANGER.)

Now it is known that the β-interaction is V-A and not S-T and consequently a reinterpretation of the RaE spectrum has become necessary. At the same time the features of electron polarization measurements raised new fundamental questions. Most of these questions are to be discussed on the basis of measurements on allowed transitions in Chap. E. Because there is a close connection between the interpretation of the RaE β-spectrum and the *polarization* of the $\beta's$ and the proof of *time-reversal invariance* from the RaE experiments, we will use results from polarization experiments discussed in Chap. D already here.

A reinterpretation of the RaE spectrum in the light of parity-non-conservation, possible failure of time reversal invariance, and a V-A law shows that the spectrum may be theoretically explained, if

$$x = i \frac{g_V}{g_A} \frac{\int \vec{r}}{\int \vec{\sigma} \times \vec{r}} \qquad (11.5)$$

and

$$y = \frac{g_V}{g_A} \frac{\int \vec{\alpha}}{\int \vec{\sigma} \times \vec{r}} \qquad (11.6)$$

lie on the curve shown in Fig. 46. The fit is only good if the phase angle between g_A and g_V deviates from $180°$ by less than approximately $6°$, and x has to lie in the interval $0.2 < x < 2.0$[3].

[1] G. ALAGA, O. KOFOED-HANSEN and A. WINTHER: Dan. Mat. fys. Medd. 28, No. 3 (1953). — J. FUJITA and M. YAMADA: Progr. Theor. Phys. Japan 10, 518 (1953). — M. E. ROSE and R. K. OSBORN: Phys. Rev. 93, 1315 (1954).
[2] E. A. PLASSMANN and L. M. LANGER: Phys. Rev. 96, 1593 (1954). — L. LIDOFSKY, N. BENCZER, P. MACKLIN and C. S. WU: Phys. Rev. 98, 1186 (1955).
[3] B. V. GESHKENBEIN, S. A. NEMIROVSKAYA and A. P. RUDIK: J. Exp. Theor. Phys. USSR. 36, 517 (1959). — JETP [translation] 9, 360 (1959).

GESHKENBEIN, NEMIROVSKAYA and RUDIK calculated the electron polarization in the case of RaE and showed that a simultaneous measurement of the energy dependence of the polarization and the β-spectrum enables a narrower limit

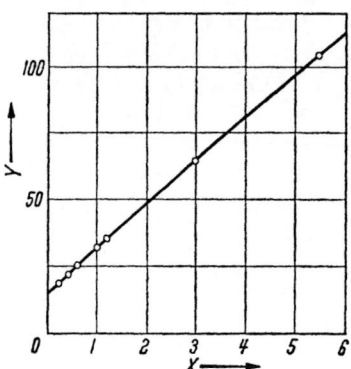

to be put on the phase of g_V/g_A and a determination of the parameter x. This procedure was followed up by ALIKHANOV, ELISEYEV and LUIBIMOV[1], who measured the electron polarization as a function of the energy. The method used in the experiment will be discussed in Sect. 14α, p. 58. The results are shown in Fig. 47. The curves shown are theoretical curves calculated under various assumptions regarding the phase between V and A [positive angle from V to A: (1) 180°; (2) 176.4°; (3) 175°]. For each phase chosen the appropriate x is chosen as the one giving the best fit to the experimental polarization results. The result is

Fig. 46. Curve showing corresponding values of x and y, as defined in Eqs. (11.5) and (11.6), optimizing the fit of the experimental RaE spectrum to the theory.

$\varphi = 176.0°\ {}^{+2.0}_{-0.5}$. The probability distribution for φ obtained from an analysis of the fit to experimental data is shown in Fig. 48.

The final result is that the RaE data can be explained if $\varphi(g_A/g_V) = 176.0°\ {}^{+2.0}_{-0.5}$; $x \sim 0.2$; $y \sim 19$. However, corrections to the theoretical picture may be relatively large. We will only regard the finding of a φ-value near 180° as supporting evidence to data found in other ways and discussed in Sect. 21.

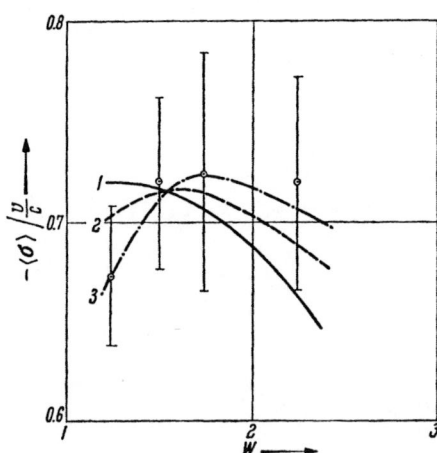

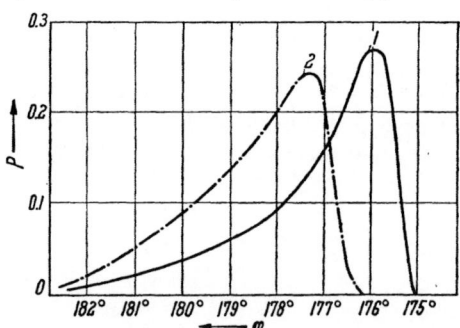

It should be noted that three entirely different theoretical spectrum assumptions were found to work in the case of the RaE spectrum. It was only with the application

Fig. 47. The results of the RaE electron polarization experiment by ALIKHANOV, ELISEYEV and LUIBIMOV. The curves are theoretical curves calculated with different values assumed for the phase of g_V/g_A and for x. (After A. J. ALIKHANOV, G. P. ELISEYEV and V. A. LUIBIMOV.)

Fig. 48. Probability distribution for the phase difference between g_V and g_A as given by ALIKHANOV, ELISEYEV and LUIBIMOV. (After A. J. ALIKHANOV, G. P. ELISEYEV and V. A. LUIBIMOV.)

of additional information that two of the possibilities could be ruled out. This is a good illustration of the difficulties usually encountered in interpretations of β-spectral shapes and discussed in Sect. 3. As a further illustration of this point, we may quote KING [5] who says in his comments on this decay that the non allowed shape does not permit good determination of β^- end-point energy.

[1] A. J. ALIKHANOV, G. P. ELISEYEV and V. A. LUIBIMOV: Nuclear Phys. 13, 541 (1959).

δ) Forbidden transitions and β-decay coupling constant determination. The determination of the coupling constants which will be discussed in Chap. E is based on experiments on superallowed transitions mostly. Some information pertaining to this point may, however, be obtained from forbidden transitions too. A particularly interesting case is the decay of RaE just mentioned. Unfortunately, most conclusions drawn from the study of forbidden β-decays are no better than our present knowledge of nuclear physics. This follows from the fact that the nuclear matrix elements, or at least nuclear matrix element ratios cannot be eliminated with sufficient precision. Thus the study of superallowed transitions, the matrix elements of which may sometimes be evaluated quite reliably, has been the most fruitful approach. Meanwhile, the improved knowledge of nuclear models gives hope that the study of forbidden transitions may gain interest in the near future.

We shall here give a brief discussion of some results which have been obtained recently from the study of forbidden transitions.

PRESTON, KEECH and PEARSON[1] have studied the β-spectrum of Rb^{87}. The decay is a third forbidden ($\Delta I = 3$, yes) $β^-$-transition with $\log_{10} ft = 17.6$. This study showed that the spectrum is consistent with a mixture of V and A interactions with the mixing ratio and the relative sign found from the neutron decay (see Sect. 21). The study was based on the nuclear shell model.

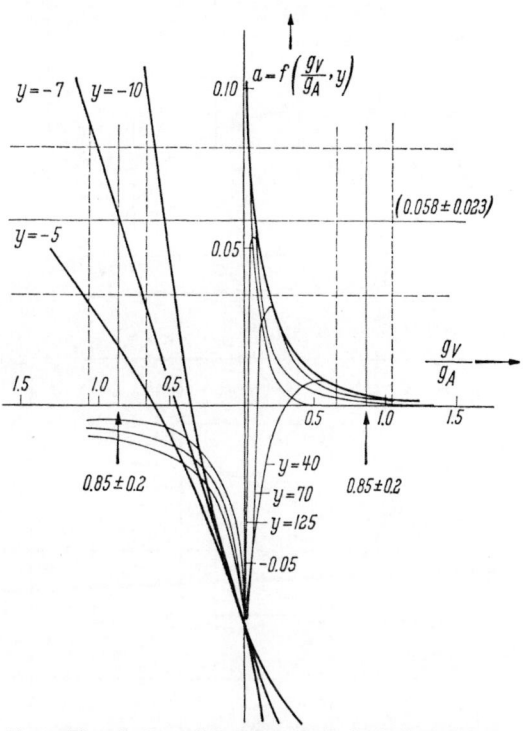

Fig. 49. The anisotropy of the β-γ-correlation in the decay of Ba^{139} as a function of the ratio g_A/g_V. The parameter y is a ratio between two matrix elements. The experimental anisotropy measured by VLASOV and RUDAKOV is also shown and so is the numerical value of g_V/g_A known from other experiments. (After N.A. VLASOV and V.P. RUDAKOV.)

Another example is the proof of time reversal invariance which was obtained from the study of RaE discussed above.

In principle the *relative sign* of the Fermi and Gamow-Teller coupling constants may be found from the study of forbidden β-spectra. This has been done by many teams. However, most of these works were performed prior to the establishment of the V and A interactions as being responsible for the β-decay process.

The relative sign of the V and A constants has been found to be negative by VLASOV and RUDAKOV[2]. They based the determination on a measurement of the angular correlation between the non-unique first forbidden 2.23 Mev β-transition of Ba^{139} and the subsequent 0.163 Mev γ-transition. The theoretical treatment requires the determination of three matrix element ratios. Two of these ratios may be obtained sufficiently well without very detailed assumptions on the nuclear model used, whereas the third ratio, called y, is more difficult to obtain. In Fig. 49 the anisotropy as a function of g_V/g_A is shown for different values of y.

[1] M.A. PRESTON, G.H. KEECH and J.M. PEARSON: Phys. Rev. **119**, 305 (1960).
[2] N.A. VLASOV and V.P. RUDAKOV: Soviet Phys. JETP **9**, 17 (1959).

The experimental value is shown with limits as well as the numerical value of g_V/g_A known from other works. It is seen that only $g_V/g_A < 0$ fits the data.

Thus it is seen that fundamental information may be obtained from the study of forbidden transitions. Further developments along these lines may be expected. But still one may accept information from (super-)allowed transitions with greater confidence.

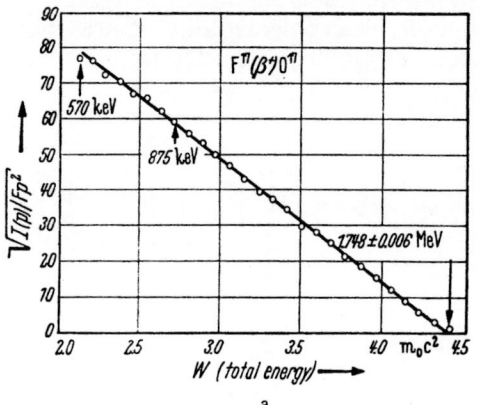

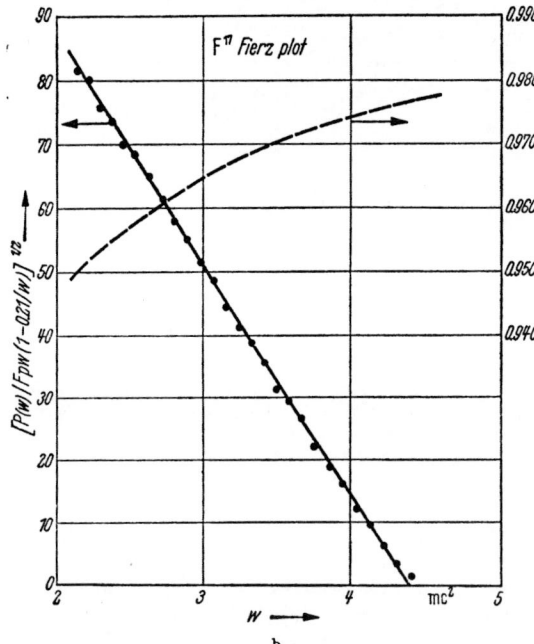

Fig. 50. a The Fermi plot for F^{17} as observed by Wong (after C. Wong); b Fierz plot of the F^{17} spectrum. Experimental points are the same as those of Fig. 50a.

12. Deviations from the Fermi shape. Before leaving the subject of spectral forms it will be convenient to touch briefly the problem of possible deviations from the statistical shape of the allowed spectra. During a long period great efforts have been invested into a search for a correction factor to the statistical shape of the form $(1+\beta/W)$. This factor was introduced in order to account for a possible interference between S and V couplings or between A and T couplings. This interference is called the *Fierz interference* [see Eqs. (6.3) and (6.4)]. The change of the spectrum due to such a factor is usually very small and consequently it is an intricate problem to decide on the existence of the Fierz factor and on the size of β.

In order to demonstrate this fact Figs. 50a and b are included. Fig. 50a is a conventional Fermi plot of the β^+-spectrum of F^{17} which is a superallowed decay, whereas b shows a Fierz plot where the experimental points are divided by $(1+\beta/W)^{\frac{1}{2}}$ where $\beta = -0.20$. This value of β is a rather alarming one requiring e.g. the numerical ratios $g_S:g_V:g_T:g_A = 0.25:1:0.25:1$. The correction factor is furthermore shown as the dotted curve in Fig. 50b. The fit to the straight line is just as good in Fig. 50b as it is in Fig. 50a. This is not at all surprising since the Fierz factor varies by less than $\pm 1.4\%$ in this case. Thus a considerable accuracy is required in this type of work.

Today it is known that because of the fact that β-particles are fully polarised, the Fierz constant β has to be very small (in effect it has to be zero, if polarization is strictly $\pm v/c$). This is independent of the relative contributions of S and V resp. T and A couplings.

Therefore the Fierz term has lost some of its importance. The results of the spectral shape examinations are that the various limits which has been put on β_{GT} fall in the range $-0.15 < \beta_{GT} < 0.1$[1]. The determination of β_F is not quite as good. Usually this constant has been determined from a single mixed transition, N^{13}. Besides these determinations the constants have been determined by other means. The most narrow limit on β_{GT} can be determined from the K capture to β^+-emission ratio (see Sect. 24). The result is $\beta_{GT} = -0.02 \pm 0.04$. β_F has been determined by GERHART[2] from an analysis of the ft values of $0 \to 0$ transitions. His result is $\beta_F = 0.00 \pm 0.12$.

Meanwhile the examination of spectral shapes has won renewed interest recently. Some examinations of possible small deviations from the statistical form of the allowed decays have shown rather queer results. For historical reasons it has become customary to fit experimental β-spectra to the Fermi spectrum multiplied by the Fierz factor $(1 + \beta/W)$ because of the earlier extensive work on Fierz interference. LANGER and coworkers[3] have made careful measurements of the β-spectra of a series of nuclides. They examined both β^--transitions (In^{114}, Y^{90}, P^{32}) and β^+-transitions (Na^{22}, Zr^{89}, Co^{56}). All these spectra could be fitted assuming $0.2 < \beta < 0.4$. If the measured deviations should be Fierz interference or outer screening effects β should show different sign for β^-- and β^+-transitions. Finite wavelength corrections are too small to explain the observed deviations. No experimental explanation can be given for the observation. As mentioned above limits less than 0.2 have been put on β from other careful spectrum measurements of e.g. He^6 (SCHWARZSCHILD et al.[1]).

Theoretical reasons for the renewed interest in the problem of the spectral shape are found in e.g. a proposal forwarded by GELL-MANN[4]. He showed that the β-spectrum should show small deviations from the statistical shape as a consequence of the assumption of a *conserved vector current* responsible for the vector interaction. In the case of a comparison of the spectra of B^{12} (β^--emitter) and N^{12} (β^+-emitter) differences of the order 20 percent should be expected.

This problem has not been tested directly by experiment[5]. But similar information may be gained from a measurement of β-γ *directional correlations*. This type of experiment has otherwise been omitted from this article. We may take the opportunity to say a few words about this point.

The β-γ correlation parameters in the expression for the angular distribution

$$W(\vartheta) = \sum_n A_n \cos^{2n} \vartheta \qquad (12.1)$$

are combinations of terms containing absolute squares of matrix elements and cross terms between matrix elements. The energy dependence of these terms may be the same or they may be different. There exists a narrow connection between these parameters and the shape factors C_n of the β-spectra. However, the cross terms contained in the A_n's and C_n's may be different. Thus it has been

[1] See e.g. J.P. DAVIDSON and D.C. PEASLEE: Phys. Rev. **91**, 1232 (1953). — A.V. POHM, R.C. WADDELL and E.N. JENSEN: Phys. Rev. **101**, 1315 (1956). — A. SCHWARZSCHILD, B.M. RUSTAD and C.S. WU: Bull. Amer. Phys. Soc., Ser. II **1**, 336 (1956). — F.T. PORTER, F. WAGNER, JR. and M.S. FREEDMAN: Phys. Rev. **107**, 135 (1957).

[2] J.B. GERHART: Phys. Rev. **109**, 897 (1958).

[3] O.E. JOHNSON, R.G. JOHNSON and L.M. LANGER: Phys. Rev. **112**, 2004 (1958). — J.H. HAMILTON, L.M. LANGER and W.G. SMITH: Phys. Rev. **112**, 2010 (1958); **119**, 722 (1960). — L.M. LANGER and J.H. HAMILTON: Bull. Amer. Phys. Soc. II **6**, 50, QA6 (1961).

[4] M. GELL-MANN: Phys. Rev. **111**, 362 (1958).

[5] *Note added in proof:* Recently T. MAYER-KUCKUK and F.C. MICHEL have compared the B^{12} and N^{12} spectra and found reasonable agreement with GELL-MAN's conserved vector current theory. See Phys. Rev. Letters **7**, 167 (1961).

found that if the β-spectrum of a given β-transition shows an allowed form, the correlation between this β-transition and a subsequent γ-transition will often be isotropic whereas non-allowed forms are often followed by an anisotropic β-γ correlation. This is not always true and due to the differences between terms in the A_n's and the C_n's additional information about matrix elements may be obtained by means of a comparison between the shape factor and the angular correlation parameters[1]. The measurement of β-γ correlations, however, presents serious problems and up till now the measurements have given little information pertaining to the topic of this article: the β-decay interaction. Thus we have omitted most of these experiments.

On the other hand, the question of deviations from the statistical shape of allowed β-spectra as asked by GELL-MANN may also be answered by means of β-γ correlation measurements. Therefore we may mention a few recent experiments. BOEHM, SOERGEL and STECH[2] have observed an anisotropic β-γ correlation in the decay of F^{20}. STEFFEN[3] has also found a significant deviation from anisotropy in the decay of Na^{22}. Quite recently PETTERSON, HAMILTON and THUN[4] have found a rather big anisotropy ($\sim 3\%$) in the high energy end of the β+-decay of Co^{56}. The correspondence with the Gell-Mann theory is not very good. This is perhaps even not to be expected. The effects are very small and may be due to other unconsidered experimental or theoretical effects. In this connection it should be remembered that LANGER and HAMILTON have shown that the Co^{56} spectrum shows a deviation from the Fermi shape as mentioned above.

As is seen the problem of the exact form of the β-spectrum has not been definitely settled, but is still an interesting experimental and theoretical problem.

D. Polarization experiments.

13. Introductory remarks. The startling revelation of parity non-conservation in β-decay has faced us with a very complex situation as regards a *systematic experimental determination of the coupling constants*. The general theory of β-decay assuming *proper Lorentz invariance* only and *no derivative couplings* leads in principle to 35 measurable and real constants[5]. Experimentally most of these constants are found to be small or probably zero. From a systematic point of view, however, it should be noted that the coupling constants usually appear in definite combinations in the measurable quantities, and an unfortunate tendency exists to combine well measurable quantities with coupling constant combinations where small or vanishing coupling constants appear as squared magnitudes thus rendering precise determinations difficult. The standard approach to this difficulty is to adopt the point of view that coupling constants are neglected if they are found to be *smaller than the main coupling constants* by an order of magnitude. Thus from the systematic experimental point of view the present approach to a determination of the coupling constants is a first-order approach only. We shall adhere strictly to this procedure in the following.

Consequently our approach is that of *eliminating all coupling constants which are found to be small*. For this purpose a discussion of the *electron polarization experiments*, the *β-γ polarization correlations*, the *angular distribution of β-par-*

[1] Cf. e.g. R.M. STEFFEN: Phys. Rev. **118**, 763 (1960).
[2] F. BOEHM, V. SOERGEL and B. STECH: Phys. Rev. Letters **1**, 77 (1958).
[3] R.M. STEFFEN: Phys. Rev. Letters **3**, 277 (1959).
[4] B.G. PETTERSON, J.H. HAMILTON and J.E. THUN: Nuclear Phys. **22**, 131 (1961).
[5] See e.g. T.D. LEE: Proc. Rehovoth Conf. Nucl. Structure 1958, p. 336.

ticles from aligned nuclei, especially the *decay of the neutron* and the *neutrino helicity* experiment will be convenient.

The various methods which can be used to measure the polarization of β- and γ-particles will not be discussed in great detail as regards the theory of the methods. For such details the reader is referred to a paper by TOLHOEK[1] on the theory of polarization, the methods of measuring electron polarization and the behaviour of electron spin in electric and magnetic fields. In a paper by SCHOPPER[2] the measurement of circular polarization of γ-rays is discussed.

14. Electron polarization. The modern theory of β-decay predicts that the β-particles should be *longitudinally polarized*. The degree of polarization should be proportional to v/c. The proportionality constant is found as a certain combination of coupling constants and the relevant matrix elements (see Sect. 20, Eq. (20.1) to (20.4)]. If parity were conserved, the proportionality constant would be zero. Therefore the polarization of β-particles was measured at an early stage of the development of the parity-non-conserving β-theory. It was shown that β-particles are really longitudinally polarized. It is of great interest to show the v/c dependence of the polarization in order to find evidence for the correctness of the theory. Furthermore an accurate determination of the proportionality constant is of importance in order to evaluate the coupling constants. The numerical value of the proportionality constant cannot be bigger than 1, in which case one talks about *full polarization*.

Many experiments have been carried out using different methods. The methods may be grouped into two groups, which are principally different. The measurement may be direct, using a *polarization dependent cross section*, e.g. Møller- and Bhabha-scattering on polarized electrons, Mott-scattering on heavy unpolarized nuclei and especially for positrons annihilation with polarized electrons or positronium formation. The measurements may also be indirect, *transferring* the longitudinal polarization of the β-particles to *circular polarization of γ-radiation* e.g. by bremsstrahlung. Then the circular polarization can be measured.

α) *Mott scattering.* The first measurement of electron polarization was carried out by FRAUENFELDER et al.[3]. The principle used in their experiment was *Mott-scattering on heavy nuclei*.

The principle of the Mott-scattering method is the following. The scattering pattern of electrons traversing a spherically symmetric electrostatic field is rotationally symmetric. In the laboratory system the electron is scattered by the electrostatic Coulomb field of the nucleus. When the system is transformed to the center of mass system the electron will feel a *magnetic field* too. This magnetic field will interact with the *magnetic moment* of the electron. The result of this interaction is that *transversely* polarized electrons will show a scattering pattern which is not symmetrical in the plane extended by the spin and momentum vectors of the incoming electrons. Thus a system of two identical detectors placed symmetrically about the above mentioned plane will show different counting rates, the difference depending on the degree of polarization.

In the electron polarization experiments the polarization to be measured is longitudinal. The Mott-scattering pattern of longitudinally polarized electrons is symmetric. Therefore some sort of *spin twisting mechanism* has to be applied to the electrons before the scattering in order to make use of the above mentioned scattering asymmetry. This can be done in several ways.

[1] H.A. TOLHOEK: Rev. Mod. Phys. **28**, 277 (1956).

[2] H. SCHOPPER: Nucl. Instrum. **3**, 158 (1958).

[3] H. FRAUENFELDER, R. BOBONE, E. VAN GOELER, N. LEVINE, H.R. LEWIS, R.N. PEACOCK, A. ROSSI and G. DE PASQALI: Phys. Rev. **106**, 386 (1957).

Frauenfelder et al. deflected the electrons in *the electric field of a cylindrical condenser*. During the deflection the electron spin will be turned only a little. By a deflection angle of 108° as used in this experiment the momentum will be turned 90° with respect to the spin. Thus the arrangement will change the longitudinal polarization to transverse polarization. The experimental arrangement is very similar to that shown in Fig. 51. The main difference is the short magnetic lens between the source and the condenser. This lens was not used in the Frauenfelder experiment. The scattering foil was a thin gold foil. Electrons scattered through an angle of from 95 to 140° were detected by means of two end-window G.M.-counters. These were arranged symmetrically about the deflection plane of the electrons. The twisting condenser acts as a momentum selector with

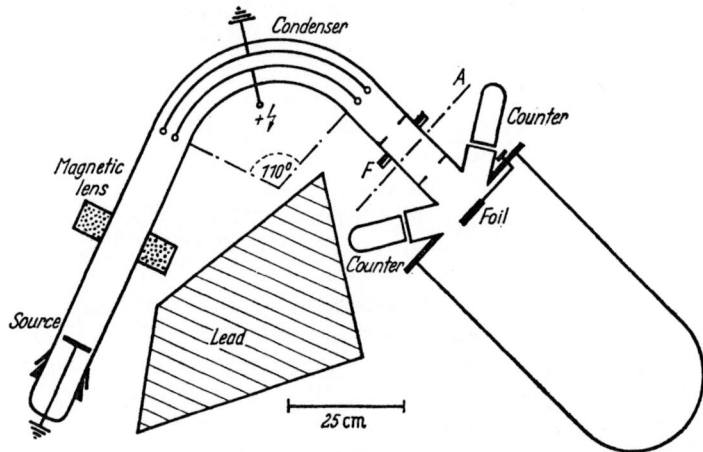

Fig. 51. The instrument used by Bienlein, Fleischmann and Wegener for the measurement of electron polarization by Mott scattering. Spin twisting is performed by deflecting the electrons in a spherical condenser. (After H. Bienlein, R. Fleischmann and H. Wegener.)

a fairly good resolution. The source must be thin in order not to cause depolarization of the electrons. A rather critical point is the scattering foil. *Plural scattering* in the foil will smear out the asymmetry. In order to prevent plural scattering the foil must be very thin. Different foil thicknesses was used in these experiments. Furthermore the gold foil was replaced by an aluminium foil. The asymmetry obtained by the Al-foil should be approximately 10% of that obtained with the Au-foil. Therefore this control experiment gives a measure of the instrumental asymmetry. In the first experiments Co^{60} was used as a source. The β-decay of Co^{60} is of the allowed G-T type ($\Delta I = 1$; no parity change). The polarization was measured for different values of the high tension applied to the condenser, that is for different values of v/c. The result -0.40 was obtained for the polarization of electrons of energy 77 kev ($v/c = +0.49$). The foil used for this point was the thinnest one used (0.05 mg/cm²). The other results are somewhat lower presumably because of *depolarization in the source* and multiple scattering in the rather thick foils used. The experiment shows that β-particles are polarized. This result is a strong indication of parity-non-conservation. The absolute value does not disagree with a $-1 \cdot v/c$ law for the polarization of β^--particles.

A rather similar experiment has been made by Bienlein, Fleischmann and Wegener[1]. Their instrument is shown in Fig. 51. The magnetic lens was included for several reasons. Better shielding against γ-radiation from the source

---

[1] H. Bienlein, R. Fleischmann and H. Wegener: Z. Physik **150**, 80 (1958).

is possible. The lens is used as an energy sorter and focussing device. This diminishes disturbing scattering in the condenser plates. The electrostatic condenser is part of a spherical condenser. The transmission of this analyser is greater than that of the cylindrical condenser by about 25 %. The chamber containing the scattering foil and the detectors is very big. A small chamber was shown to give too much disturbing back-scattering. The deflection angle was 110° and the scattering angle about 120°, where the asymmetry is maximum. Scattering foils of Al, Ag and Au were used. The foils were made by evaporation (50 to 800 μg per cm²) on thin plastic foils (40 μg/cm²). The authers claim that for the thin foils multiple scattering effects can be neglected. A correction for source backing effects was calculated. In their first experiment Co⁶⁰ was used. The electron

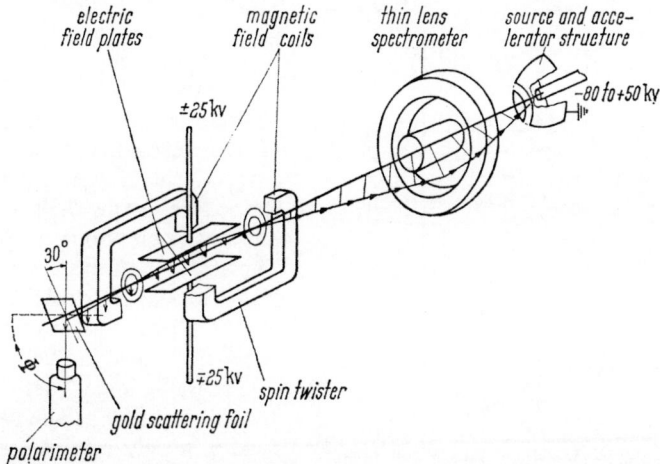

Fig. 52. The instrument used by CAVANAGH et al. for the measurement of electron polarization by Mott scattering. The spin twisting is performed during the passage through the crossed B and H fields. (After P.E. CAVANAGH, J.F. TURNER, C.F. COLEMAN, G.A. GARD and B.W. RIDLEY.)

energy 166 kev was chosen ($v/c = 0.66$). This particular energy was chosen because a β-spectrum measured by means of the thin lens and the condenser deviated from the statistical shape at lower energies. The result of the measurement was -0.635 ± 0.05. The result may be raised by about 5 % when the Mott-scattering cross section is corrected for the *screening by the electron cloud* of the Coulomb field.

It is possible to perform the spin twisting by transmission through other electromagnetic fields. CAVANAGH et al.[1] and ALIKHANOV et al.[2] have used *crossed electric and magnetic fields*. The ratio B/H defines a certain electron energy where the electrons go undeflected through the crossed fields. The field will turn the spin. By a suitable choice of B against the length of the field region the longitudinal polarization will be changed to transverse polarization. The instrument used by CAVANAGH et al. is shown in Fig. 52. A thin magnetic lens selects electrons of a certain energy. The electrons are focussed onto the opening of the B-H field. The reasons for using the lens instead of relying on the selectivity of the B-H field are the same as discussed above in connection with Fig. 51. The scattering foil was again a gold foil. The foil was tilted so as to measure always in transmission. The effect of plural scattering is minimized in this way. The scattered β-particles were detected by one plastic scintillation counter in con-

[1] P.E. CAVANAGH, I.F. TURNER, C.F. COLEMAN, G.A. GARD and B.W. RIDLEY: Phil. Mag., Ser. VIII **2**, 1105 (1957).
[2] A.J. ALIKHANOV, G.P. ELISEYEV, V.A. LUIBIMOV and B.V. ERSHLER: Nuclear Phys. **5**, 588 (1958).

nection with a pulse height analyzer. The scatterer-detector arrangement could be rotated in order to measure the angular distribution as a function of the azimuthal angle. Control experiments with Al-foils were made and different foil thicknesses were used. The foils were made by evaporation of 100 to 1000 μg per cm² material on 20 μg/cm² nylon films. Again Co⁶⁰ was used for the first experiment. The source was better than 1 mg/cm² on 1.2 mg/cm² aluminium. The polarization of 128 kev electrons was measured with the result − 0.65 ±0.13. This agrees with the v/c value which is 0.6 at this energy.

Later on a beautiful modification of the source arrangement was included in the instrument[1] in order to improve the measurement of the energy dependence of the polarization. In fact the instrument shown in Fig. 52 is this modified version.

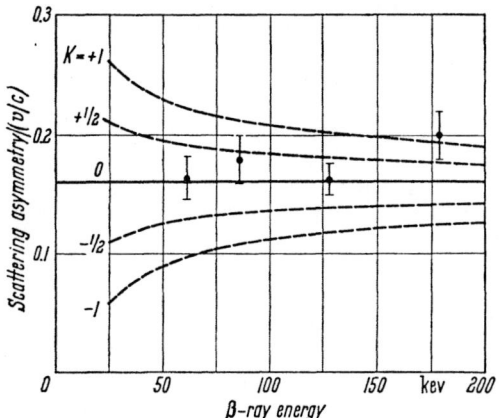

Fig. 53. Energy dependence of the polarization of Co⁶⁰ β⁻-particles as measured by Cavanagh et al. (After P.E. Cavanagh.)

A potential was applied between the source and the body of the instrument. The electrons are pre-accelerated in this potential difference. In order not to change the polarization of the electrons a suitably shaped earthed shield was arranged around the source. A ring drilled in this shield allows the electrons, to pass through. The shield is shaped so as to *minimize the transverse electrostatic forces* acting on the electrons traversing the aperture. As is well known a longitudinal electrostatic force does not influence the polarization of the particles if the field is purely electrostatic. Thus all electrons leaving the potential region with an energy E has a polarization as that of a β-particle of energy $E − V$ where V is the potential of the source with respect to the body of the instrument. In order to measure the energy dependence of the polarization all parameters but the potential were kept constant. Thus the energy of the electrons selected by the original part of the instrument was always 128 kev. By varying the potential between + 70 kev and − 50 kev the polarization of the β-particles was measured in the energy interval from 58 to 178 kev covering a range of v/c values from 0.45 to 0.70. The results obtained for Co⁶⁰ are shown in Fig. 53. The curves are theoretical ones calculated under the assumption

$$P/(v/c) = 1 + k \frac{\alpha Z}{p} \tag{14.1}$$

adopting different values of k. The second term in Eq. (14.1) is a Coulomb correction term of usual type. k depends on the coupling constants. The results of Fig. 53 are preliminary. They show essentially the linear v/c dependence although the statistics are not good enough for a decision on the accurate value of k. The advantage of this method is of course that many errors which can not be corrected for may cancel out because the apparatus can work at constant electron energy.

The experiment made by Alikhanov et al. (p. 55, Ref. 2) was very similar. The magnetic lens was omitted and the B-H field was used as energy analyzer. The counters used were two G.M. counters placed behind each other. The counters were separated by an absorber acting as energy discriminator, and coincidence was required. In this experiment Sr⁹⁰ + Y⁹⁰ was examined. All transitions of

[1] P.E. Cavanagh: Proc. Roy. Soc. Lond. **246**, 466 (1958).

interest in this source are unique first forbidden. The same can be said about an impurity in the source, Sr^{89}. Electrons of energy 300 and 750 kev were examined. The scattering foil was 0.5 mg/cm² gold. The results were corrected for finite scatterer thickness and were $(-1.02 \pm 0.15)\ v/c$ and $(-1.15 \pm 0.4)\ v/c$ respectively.

In the experiments hitherto discussed deflection in macroscopic fields have been used as spin twisting mechanism. DE SHALIT et al.[1] used microscopic fields. Their instrument is shown in Fig. 54. The spin twisting is here accomplished by 90° *Mott-scattering*. During the deflection the spin is also changed a little but after the deflection the polarization will have a large transverse component.

The geometry of the spin twisting material is chosen semicircular. The source and the analyzing foil extends a diameter on the circle. Thus all electrons are scattered through 90°. The analyzing scattering angle is approximately 90°. The spin twisting foil was 125 mg/cm² aluminium and the analyzing foil was 2.5 mg/cm² gold. The detector arrangement consisted of two scintillation counters with pulse height discriminators setting lower limits on the energy of the elec-

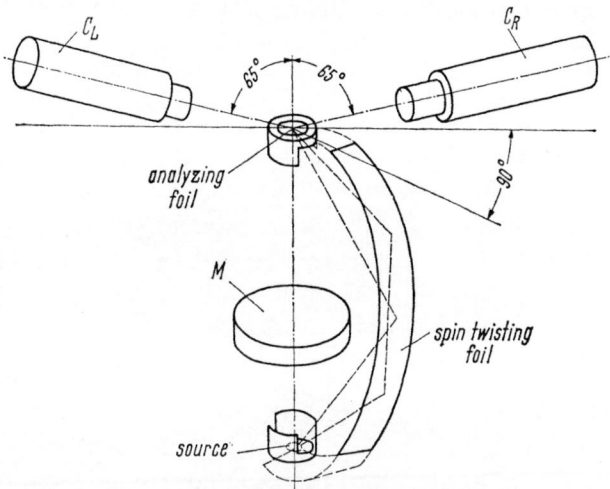

Fig. 54. The instrument used by DE SHALIT et al. for the measurement of electron polarization by Mott scattering (double scattering method). The spin twisting is performed by 90° scattering in an Al foil. (After A. DE SHALIT.)

trons recorded. An effective upper limit was set by the $1/E^2$ dependence of the Mott-scattering cross section on energy. In the first experiment P^{32} was used. The result showed a definitive backwards polarization of the electrons. The result has not been calculated and compared with theory quantitatively. A relative measurement of the polarization of electrons from P^{32} and Au^{198} has been made with comparable results.

One might expect that plural scattering in the foils could disturb the measurements. This is really the case if the analyzing foil is not very thin. This difficulty can be overcome by repeating the experiments using different foil thicknesses and extrapolating the results to zero foil thickness. This procedure has been used by many authors. The twisting foil need not be very thin. The reason is that in effect the plural scattering process acts as a spin twisting process. The degree of transverse polarization after a deflection in a cylindrical field is essentially the same as that obtained by multiple scattering through the same angle. J. HEINTZE has mentioned this fact in connection with his experiments on double scattering of β-particles (see Table 4, Ref. 1).

Also ALIKHANOV, ELISEYEV and LUIBIMOV[2] have used the double scattering method. The instrument is shown in Fig. 55. The double scattering method was chosen in order to make a simple and reliable instrument. The idea was to measure

[1] A. DE SHALIT, S. KUPERMAN, H. J. LIPKIN and T. ROTHEM: Phys. Rev. **107**, 1459 (1957); **109**, 223 (1958).

[2] A. J. ALIKHANOV, G. P. ELISEYEV and V. A. LUIBIMOV: Nucl. Phys. **7**, 655 (1958).

the energy dependence of the polarization of electrons from many different sources. The principle of the instrument needs no explanation beeing the same as that underlying the instrument used by DE Shalit et al. The spin was changed to transverse by 90° scattering in a rather thick foil. The twisting characteristics of this foil were calculated by a Monte Carlo calculation taking multiple scattering into account. Each of the two detectors consisted of a pair of G.M. counters separated by an energy discriminating absorber and run in coincidence. Many control experiments were carried out in order to allow corrections for different effects. Two different instruments were used. They were identical except for the choice of materials. The use of these instruments made possible the evaluation of the corrections for backscattering from the walls of the instrument and from the

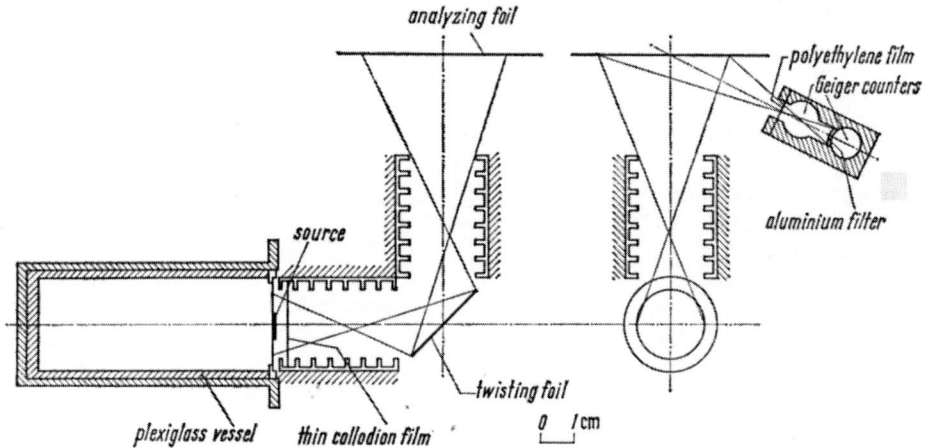

Fig. 55. The instrument used by Alikhanov, Eliseyev and Luibimov for the measurement of electron polarization by Mott scattering (double scattering). (After A. J. Alikhanov, G. P. Eliseyev and V. A. Luibimov.)

source support. Also the influence of electrons initiated by scattering of γ-radiation was examined. Different materials were shown to be equally good as spin twisting materials. The energy distribution of the electrons leaving the spin twisting foil was measured by means of a spectrometer. This was imperative because the energy selection of the instrument is determined by the spin twisting foil and the absorber between the two coincident counters. Different thicknesses of the analyzing foil were used and the results extrapolated to zero thickness. All the effects mentioned were included in the calculations and furthermore corrections were applied for effects caused by conversion electrons from the source, solid angles of the different parts of the instrument, shielding of the Coulomb field of the nuclei (i.e. correction of the Mott scattering cross section), and depolarization in the source. The sources used were all first forbidden β-transitions with parity shift: Tm^{170} $(\Delta I=1)$, Re^{186} $(\Delta I=1)$, Sm^{153} $(\Delta I=1$ and $0)$, Au^{198} $(\Delta I=0)$, Lu^{177} $(\Delta I=1$ and $0)$ and for comparison $Sr^{90}-Y^{90}$ $(\Delta I=2,$ unique$)$. All of the transitions are mixed Fermi and Gamow-Teller transitions. For each of the sources different energy settings, that is different thicknesses of spin twisting and absorber foils, were used. The results of the measurements are shown in Fig. 56. All numbers $(-P/(v/c))$ are identical to within 10% or better. The polarization is proportional to v/c to within 4 to 7% and the proportionality constant (calculated as the mean of all the measurements) is -1 ± 0.03.

The above discussion represents only some of the Mott scattering experiments which have been performed. They give a fairly good representation of the different possibilities which are at the diposal of the experimentalist. For a survey of the results mentioned and some other results see Table 4, p. 66 to 69.

β) Møller and Bhabha scattering. Another method which has been used extensively to measure electron polarization is electron-electron scattering (Møller scattering) or positron-electron scattering (Bhabha scattering). The cross section for scattering of a polarized β-particle on an electron depends on the relative orientation of the spins of the two particles. Thus if the β-particles from a source are scattered in a suitably oriented magnetized iron foil, a detector covering a certain solid angle should show different counting rates for opposite magnetization directions. The most serious problem which one has to solve in order to carry out this experiment is that the detector has to discriminate a few electron scattering events from an overwhelming background of Rutherford scattering events. This problem is best tackled by imposing a coincidence requirement. The atomic electron participating in the electron scattering event receives an appreciable part of the energy of the incoming electron or positron and could thus be detected along with the primary particle. Two electron detectors recording coincident pairs of electrons do not record the Rutherford scattered electrons and consequently such an arrangement has the required

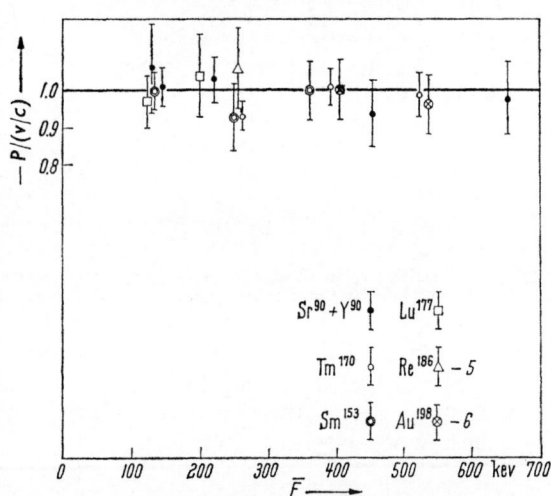

Fig. 56. The energy dependence of $-P/(v/c)$ for different nuclides as obtained by ALIKHANOV et al. using the instrument shown in Fig. 55. (After A. J. ALIKHANOV, G. P. ELISEYEV and V. A. LUIBIMOV.)

selectivity. The electron-electron scattering method has the advantage over the Mott scattering method that the former do not require any spin twisting.

Polarization experiments using electron-electron scattering have been carried out by many authors, for instance by FRAUENFELDER et al.[1], BENCZER-KOLLER, SCHWARZSCHILD, VISE and WU[2] and by GEIGER, EWAN, GRAHAM and MACKENZIE[3]. The instrument used by GEIGER et al.[3] is shown in Fig. 57. A beam of β-particles is defined by apertures. The divergence of the beam is 8°. A magnetized Deltamax foil is tilted 30.5° to the beam direction. This angle which is in effect the angle between the directions of polarization of the two colliding particles has to be kept small in order to maximize the asymmetry effect. The two detectors are plastic scintillation counters placed symmetrically about the plane subtended by the two spin directions. Both detectors subtend an angle from 28.5 to 46.5°. The symmetrical arrangement corresponds to equal sharing of the energy between the two particles. The scattering angle is chosen so as to maximize the asymmetry (see Fig. 58). Single channel analyzers select pulses of correct amplitudes and a fast-slow coincidence analyzer is used. A rather thick

[1] H. FRAUENFELDER, A. O. HANSON, N. LEVINE, A. ROSSI and G. DE PASQALI: Phys. Rev. **107**, 643 (1957).

[2] N. BENCZER-KOLLER, A. SCHWARZSCHILD, J. B. VISE and C. S. WU: Phys. Rev. **109**, 85 (1958).

[3] J. S. GEIGER, G. T. EWAN, R. L. GRAHAM and D. R. MACKENZIE: Phys. Rev. **112**, 1684 (1958).

Deltamax foil was used (2.65 mg/cm²). Plural scattering in the foil was corrected for and of course finite energy resolution and finite solid angle effects were considered

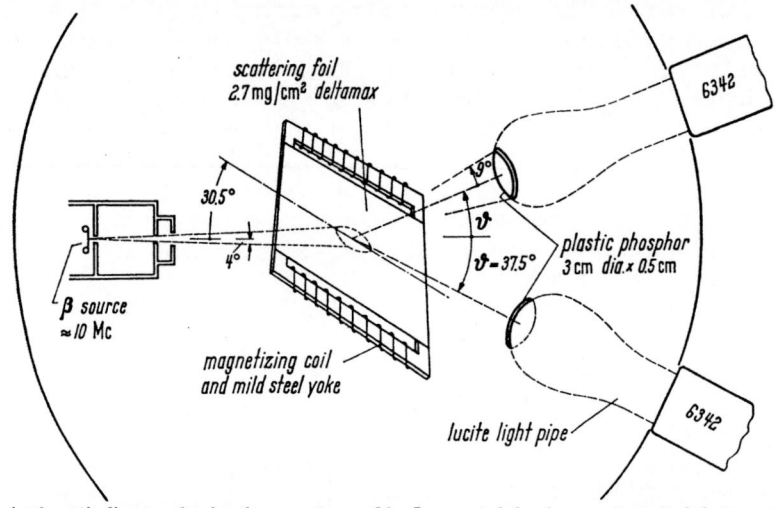

Fig. 57. A schematic diagram showing the apparatus used by Geiger et al. for the measurement of electron polarization by Møller scattering. (After J.S. Geiger, G.T. Ewan, R.L. Graham and D.R. Mackenzie.)

too. Geiger et al. examined Au^{198} ($\Delta I=0$, yes), P^{32} ($\Delta I=1$, no), Y^{90} ($\Delta I=2$, yes), Pr^{144} ($\Delta I=1$, yes), and RaE ($\Delta I=1$, yes). The results as shown in Fig. 59 are in reasonably good agreement with a polarization $-v/c$ except that for RaE which is discussed separately in Sect. 11γ.

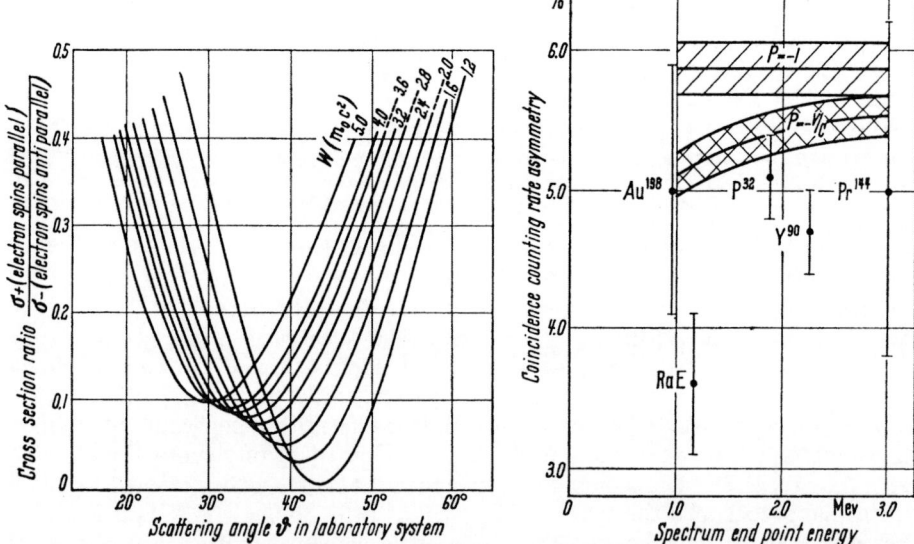

Fig. 58. Cross section for Møller scattering as a function of ϑ and W. (After J.S. Geiger, G.T. Ewan, R.L. Graham and D.R. Mackenzie.)

Fig. 59. Counting asymmetries as measured by Geiger et al. with the apparatus shown in Fig. 57. (After J.S. Geiger, G.T. Ewan, R.L. Graham and D.R. Mackenzie.)

γ) *Annihilation radiation.* The methods hitherto discussed have been used almost exclusively for measurement of *electron polarization.* In positron experiments the unavoidable *annihilation radiation* will always introduce extra difficulties. But due to some special characteristics of annihilation radiation the radiation itself may be used for *positron polarization* measurement.

When a positron leaves a nucleus its fate is usually the following. Gradually it looses its energy by scattering and bremsstrahlung. Occasionally during the slowing down a positron is annihilated together with an electron of the stopping matter. In this case one talks about *annihilation-in-flight*. Only a few percent of the positrons are annihilated-in-flight during the slowing down, the slowing down process being very fast. The positrons succeeding in being *thermalized* are annihilated within a period ranging from 10^{-10} sec (in solid matter) to 10^{-7} sec (in gases). In this case one may still talk about annihilation-in-flight the atomic electrons participating in the process having energies of some electron volts. In a certain energy interval during the slowing down there is a chance that the positron may capture an atomic electron and form an atomlike configuration called *positronium*. The positronium is formed in the *singlet* or *triplet states* with

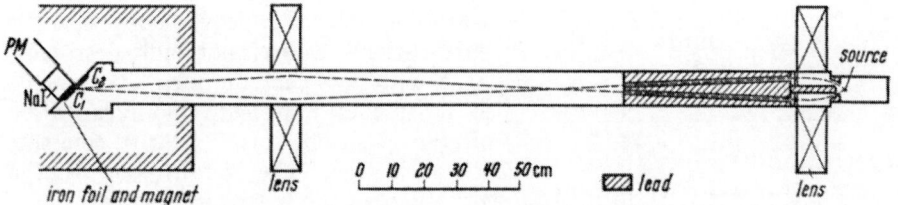

Fig. 60. The apparatus used by FRANKEL et al. for the measurement of the positron polarization by annihilation-in-flight. (After S. FRANKEL, P. G. HANSEN, O. NATHAN and G. M. TEMMER.)

the statistical weights $\frac{1}{4}$ and $\frac{3}{4}$, respectively. The singlet state decays by two-quantum annihilation with a lifetime of approximately 10^{-10} sec. The triplet state decays by three-quantum annihilation with a lifetime of approximately 10^{-7} sec. This is an oversimplified and incomplete picture of the annihilation of positrons. Especially the fate of once formed positronium is very complex.

These processes may be used to measure positron polarization in different ways. A common difficulty in all these experiments is that the polarization may be destroyed to some extent during the slowing down. Thus for good absolute measurements annihilation-in-flight near the primary energy ought to be used. Even in this case the experiments are extremely difficult.

FRANKEL, HANSEN, TEMMER and NATHAN[1] have measured the annihilation-in-flight rate for positrons traversing a magnetized iron foil. This rate will be different for two opposite magnetization directions. This is a consequence of the fact that due to the symmetry properties of the two-quantum system, two-quantum annihilation is forbidden when the two particles occupy a ^3S-state. Furthermore annihilation from higher L-states is very improbable due to the "great distance" between the particles. Consequently the spins must be antiparallel. In magnetized iron there is a preferred electron-spin direction. Thus the annihilation rate of positrons in magnetized iron must be greater when the polarization direction and the preferred spin direction of the electrons are opposite and less in the opposite case. The instrument used by FRANKEL et al. is shown in Fig. 60. A thin magnetic lens selects positrons of a certain energy and provides space for shielding against γ-radiation from the source. A second lens selects the positrons from the heavy annihilation radiation produced in the walls of the lead shielding of the first lens. The positrons hit a magnetized iron sheath. Because of the

[1] S. FRANKEL, P. G. HANSEN, O. NATHAN and G. M. TEMMER: Phys. Rev. **108**, 1099 (1957).

momentum of the positrons the annihilation radiation will proceed preferentially in the forward direction. The quanta are detected by a 3×3 in. NaI(Tl) scintillation detector placed immediately behind the iron sheath. Only photons of energy higher than 0.5 Mev are counted, i.e. annihilation of thermalized positrons is excluded. Due to the fact that the iron sheath is very thick the annihilation peak in the forward direction will be smeared out by multiple scattering of the positrons. Consequently the narrow geometry is preferable. The calculation of the effect to be expected is very difficult. The conclusion to be drawn from this experiment is that positrons from Ga^{66} $(0^+ \rightarrow 0^+)$ are highly polarized. The polarization is parallel to the momentum.

As mentioned above one might expect that the slowing down process would destroy the polarization. The treatment of this problem is very complicated as is the slowing down process itself. An important role is played by Coulomb scattering. In a first approximation the spin is left unchanged during a Coulomb scattering. This fact leaves hope that the polarization is at least not badly destroyed. As a matter of fact the next experiments to be discussed measure the polarization of the positrons after nearly complete thermalization. The results show that a great part of the polarization is left.

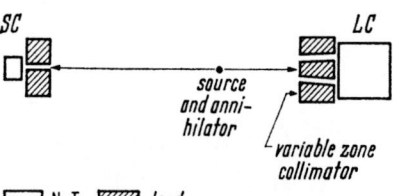

Fig. 61. Schematic diagram of the instrument used by Hanna and Preston for the measurement of the positron polarization by the study of the angular correlation between γ-quanta from annihilation-in-rest in magnetized iron.

Hanna and Preston[1] have studied the angular correlation of the two photons from a two-quantum annihilation. As previously mentioned there is a marked difference between the annihilation cross sections with spins parallel or antiparallel. Consider the case of annihilation in a big iron sample placed in a magnetic field. Most of the positrons are thermalized before annihilating. Thus the angular distribution of the annihilation quanta is a measure of the momentum distribution of the electrons participating in the annihilation processes. Because of the spin selection rule for the annihilation the angular distribution will be determined by the energy distribution of the electrons having spin antiparallel to the positron spin if the positron polarization is complete. The effect of the magnetic field applied to the iron sample will be that the valence electrons are aligned. Thus the part of the correlation curve corresponding to the momentum of the valence electrons will be higher for the field direction which aligns the spin suitably, that is when the magnetic field direction and the positron spin direction are parallel.

The instrument used by Hanna and Preston is shown schematically in Fig. 61. The arrangement is rotationally symmetric. The annihilating sample is placed between two NaI(Tl) scintillation counters. One of the counters is collimated by means of a circular hole in a lead block. The solid angle defined by this hole is very small. The apex angle is 3.3 milliradians. The other NaI counter is collimated by means of a ring shaped aperture determining the angle between the two gamma quanta. The ring collimator is made up of a conical lead cylinder and a conical ring. The ring and the cylinder can be replaced by other rings and cylinders in order to vary the angle between the two detected photons and the angular resolution of the collimator. This arrangement permits a measurement of a differential or integral correlation. An example of the differential correlation is shown in Fig. 62 for two different directions of the magnetic field. The effect of the magnetic field is clearly seen. The quantitative interpretation of the experimental results

[1] S. S. Hanna and R. S. Preston: Phys. Rev. 106, 1363 (1957); 108, 160 (1957); 109, 716 (1958).

is difficult. In order to interprete the result the momentum distribution of the electrons participating in the annihilation has to be known. And even then the polarization which is measured is the polarization at the moment of annihilation and not the original polarization of the positrons. But the experiments show very clearly that the positrons are polarized parallel to the momentum to a high degree. HANNA and PRESTON have examined the pure G.T. transition in Cu^{64} and the mixed transition in N^{13} with comparable results.

PAGE and HEINBERG[1] have used positronium formation for polarization measurements. Again the experiment consists of a measurement of the angular correlation between the two annihilation quanta. The experimental arrangement is very similar to that shown in Fig. 61. The iron annihilator shown is replaced by a gas sample chosen so as to favour the positronium formation. The collimators are replaced by line shaped apertures and the angle is changed by moving one of the counters. The angular correlation is different for two different directions of the magnetic field if the positrons are polarized. The idea underlying this experiment is totally different from that of the experiment of HANNA and PRESTON. In the annihilation process the annihilation from positronium states with $L > 0$ can be neglected due to the fact that these states will decay to the S-state before annihilating. The S-state can have $I = 0$ or 1 (singlet or triplet respectively). The singlet state will decay by two-quantum annihilation with a half life of the order 10^{-10} sec. Selection rules forbid the two-quantum annihilation of the triplet state. This state decays through three-quantum annihilation with a lifetime of the order 10^{-7} sec. By application of a magnetic field to the positronium the eigenstates will be modified. The original triplet state will be mixed up with a certain amount of the singlet state and vice versa. In the field free case the singlet and triplet states are mixtures of equal amounts of the two components: positron spin up-electron spin down and positron spin down-electron spin up. The mixing of the original singlet and triplet state works so that the new singlet state contains more of the component positron spin up-electron spin down, i.e. both magnetic moments up in the case of a magnetic field applied in the up-direction, whereas the new triplet state contains most of the component both magnetic moments down. In the case of high field the angular moments are completely decoupled and the singlet state will be the pure state both magnetic moments up. In the case of a small magnetic field the new triplet state will be able to decay weakly by two-quantum annihilation through the admixture of the singlet state. As long as the fields are weak the life time of the triplet will be essentially longer than that of the singlet. Due to these facts it is possible to measure separately the formation rates of triplet states and singlet states. This

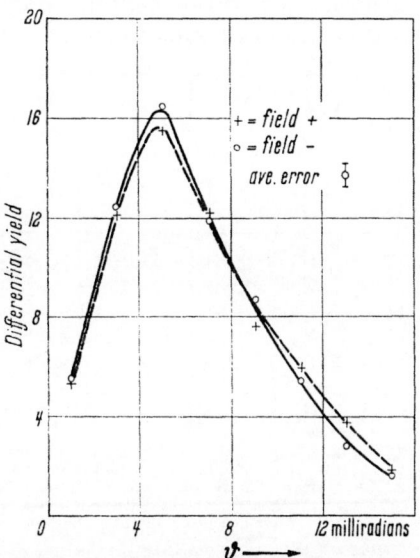

Fig. 62. An example of the differential angular correlation between the two annihilation quanta originating from positron annihilation in magnetized iron as measured by HANNA and PRESTON. The two curves correspond to two different magnetic field directions. (After S.S. HANNA and R.S. PRESTON.)

[1] L.A. PAGE and M. HEINBERG: Phys. Rev. 106, 1220 (1957).

measurement requires a separation between two-quantum annihilation from triplet and singlet states. This is possible considering the different life times of the two states. The gas sample, in which the positronium is formed and decays, could be so chosen, that the time required to thermalize the positronium by scattering is intermediate between the half lives of the two states. In this case the angular correlation between the two quanta from a singlet state will be broader than that from the triplet state, because the singlet state decays so fast that very little energy is lost by scattering while the triplet state is almost completely thermalized before annihilation. Thus it is possible to measure the relative formation rates of singlet and triplet positronium by analyzing the angular correlation. In order to make use of this fact one more effect of the magnetic field has to be considered.

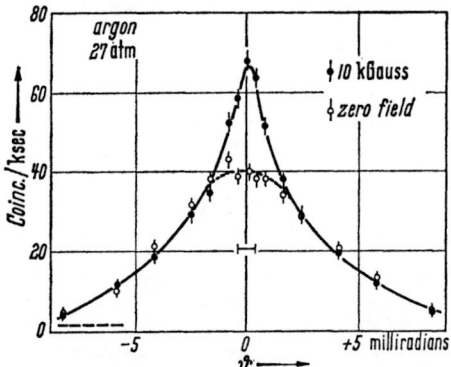

Fig. 63. An example of the angular correlation between the γ-quanta from two-quantum annihilation of positronium as measured by Page and Heinberg. The dotted curve represents singlet annihilation at zero magnetic field. (After L. A. Page and M. Heinberg.)

As mentioned above the component positron spin up (that is both magnetic moments up) dominates the singlet state in the case of a magnetic field in the up-direction. The component with positron spin down dominates the triplet state. Thus if magnetic field and positron spin are parallel, the singlet state formation will be enhanced against that of the triplet state. Likewise the singlet state formation is depressed if the magnetic field and the positron spin are anti-parallel. The amount of the two-quantum annihilation of the triplet state induced by the magnetic field is the same for both field directions. Applying these arguments to the actual measurement of the angular correlation it is seen that the difference between the narrow components measured for two directions of the magnetic field is a measure of the positron polarization. Fig. 63 shows an example of the angular correlation. The dotted curve represents the broad singlet annihilation measured at zero field. The narrow component is clearly seen. The result of this experiment is that Na^{22} positrons $(3^+ - 2^+$, allowed G.T.) are polarized in the direction of the momentum to a degree higher than $+0.4 \cdot v/c$ at the moment of annihilation.

δ) *Indirect methods.* A common characteristic of the experiments hitherto discussed is that the measurement is direct, using a polarization dependent cross section. Also more indirect methods have been used. In some reactions involving polarized electrons or positrons some secondary particle may be polarized to a degree depending on the polarization of the electron. A measurement of the polarization of this secondary particle may then determine the electron polarization.

To this group of experiments belong for positron polarization the experiments carried out by Deutsch et al.[1] and Boehm et al.[2]. These experiments are very similar. The principle is the following. If a longitudinally polarized positron is annihilated, the annihilation quanta will be circularly polarized to a degree depending on the degree of polarization of the positrons. Thus by measuring the circular polarization of the annihilation quanta information on the positron

[1] M. Deutsch, B. Gittelman, R. W. Bauer, L. Grodzins and A. W. Sunyar: Phys. Rev. 107, 1733 (1957).
[2] F. Boehm, T. B. Novey, C. A. Barnes and B. Stech: Phys. Rev. 108, 1497 (1957).

polarization is obtained. The arrangement used by DEUTSCH et al. is shown in Fig. 64. Positrons of the desired energy are selected in a thin magnetic lens and focussed on a Lucite converter. The circular polarization of the gamma quanta is measured by a transmission experiment the principle of which is the following. The Compton scattering cross section depends strongly on the orientation of the photon momentum and po-

larization directions relative to that of the electron spin. If iron is used as an absorber some of the electrons can be polarized by a magnetic field. This effect will cause a difference in absorption rate for two opposite magnetization directions. The difference is a measure of the circular polarization of the gamma quanta. The best result of the Deutsch experi-

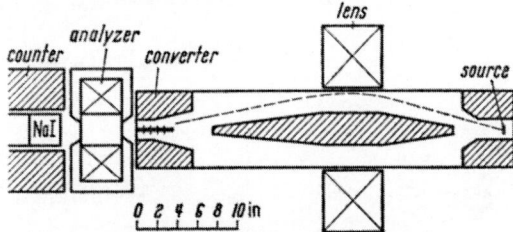

Fig. 64. The instrument used by DEUTSCH et al. for the measurement of the positron polarization by measuring circular polarization of annihilation quanta. (After M. DEUTSCH, B. GITTELMAN, R.W. BAUER, L. GRODZINS and A.W. SUNYAR.)

ments was obtained with Ga⁶⁶, the results indicating a high degree of positive polarization. An experiment on Cl³⁴ lead to the same conclusion.

Also bremsstrahlung has been used to convert electron polarization to circular polarization of gamma radiation. GOLDHABER, GRODZINS and SUNYAR[1] measured

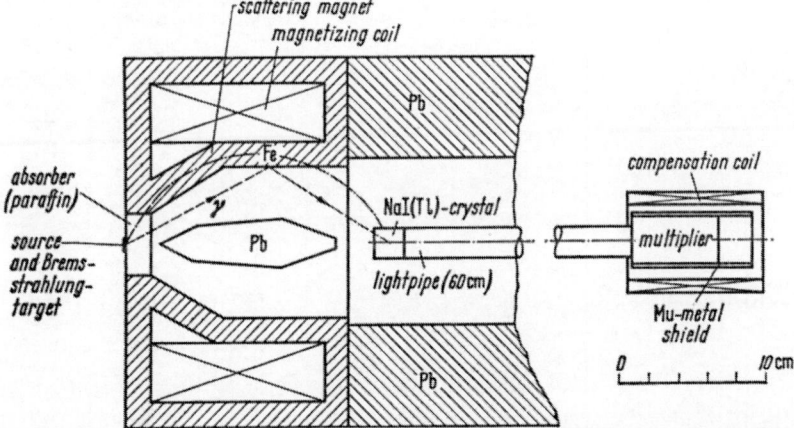

Fig. 65. The instrument used by SCHOPPER and GALSTER for the measurement of electron polarization. The circular polarization of bremsstrahlung is measured by forward Compton scattering in magnetized iron. (After H. SCHOPPER and S. GALSTER.)

gamma polarization by the transmission method discussed above. SCHOPPER and GALSTER[2] measured the gamma polarization by a different method. Still Compton scattering was used. Instead of measuring the absorption rate the rate of gamma quanta scattered through a certain angle by magnetized iron was measured for two directions of magnetization. The experimental arrangement is shown in Fig. 65. Bremsstrahlung is produced in a lead absorber which has a high bremsstrahlung cross section. The bremsstrahlung is scattered in the iron cylinder shown and afterwards detected by a NaI(Tl) scintillation counter. The theoretical calculation is very cumbersome. The method suffers from the same fault as mentioned before that the electrons are completely stopped in the

[1] M. GOLDHABER, L. GRODZINS and A.W. SUNYAR: Phys. Rev. 106, 826 (1957).

[2] H. SCHOPPER and S. GALSTER: Nuclear Phys. 6, 125 (1958).

Table 4.

Decay	Character	E_{Max} kev	Electron E kev	$\langle v/c \rangle$
$P^{32} \xrightarrow{\beta^-} S^{32}$ $(1^+ \to 0^+)$	Allowed, pure G-T	1710	300 to 1000	0.85
			800 to 1600	0.94
			200 to 500	∼0.80
			520 to 710	0.8915
			168 ± 5%	0.66
			>300	>0.78
$Co^{60} \xrightarrow{\beta^-} Ni^{60}$ $(5^+ \to 4^+)$	allowed, pure G-T	310	128 ± 5%	0.6
			different	different
				0.49
			166 ± 5%	0.66
			168 ± 5%	0.66
$Sc^{46} \xrightarrow{\beta^-} Ti^{46}$ $(4^+ \to 4^+)$	allowed, mixed	360	168	0.66
$Y^{90} \xrightarrow{\beta^-} Zr^{90}$ $(2^- \to 0^+)$	unique, first forbidden, pure G-T	2300	600 to 900	0.96
			>300	>0.78
$Sr^{90} \xrightarrow{\beta^-} Y^{90} \xrightarrow{\beta^-} Zr^{90}$ $(0^+ \to 2^- \to 0^+)$	unique, first forbidden, pure G-T	500 and 2300	different	different
			200 to 400	0.8
			∼300	0.78
			∼700	0.92
$Au^{198} \xrightarrow{\beta^-} Hg^{198}$ $(2^- \to 2^+)$	first forbidden	960	128	0.6
			170	0.66
			>600	0.9
			>300	0.9
			200 to 600	0.8
			different	different
$Tl^{204} \xrightarrow{\beta^-} Pb^{204}$ $(2^- \to 0^+)$	unique, first forbidden	760	200 to 600	0.8
$Tm^{170} \xrightarrow{\beta^-} Yb^{170} \left(1^- \to \begin{smallmatrix} 0^+ \\ 2^+ \end{smallmatrix}\right)$	first forbidden	968 884	170	0.66
$Re^{186} \xrightarrow{\beta^-} Os^{186} \left(1^- \to \begin{smallmatrix} 0^+ \\ 2^+ \end{smallmatrix}\right)$	first forbidden	1072 934	different	different
$Sm^{153} \xrightarrow{\beta^-} Eu^{153} \left(3/2^- \to \begin{smallmatrix} 5/2^+ \\ 3/2^+ \\ 5/2^+ \end{smallmatrix}\right)$	first forbidden	803 698 640	different	different
$Lu^{177} \xrightarrow{\beta^-} Hf^{177} \left(7/2^+ \to \begin{smallmatrix} 9/2^- \\ 1/2^- \end{smallmatrix}\right)$	first forbidden	380 497	different	different
$Y^{91} \xrightarrow{\beta^-} Zr^{91}$ $(1/2^- \to 5/2^+)$	unique, first forbidden	1540	200 to 600	0.8
$Ho^{166} \xrightarrow{\beta^-} Er^{166} \left(0^- \to \begin{smallmatrix} 0^+ \\ 2^+ \end{smallmatrix}\right)$	first forbidden	1840 1760	200 to 600	∼0.8

Table 4.

$\dfrac{P}{v/c}$	Method	Reference
$-1{,}00 \pm 0.13$ $-1{,}00 \pm 0.17$	} Møller scattering	FRAUENFELDER [3]
high negative	double scattering	DE SHALIT, p. 57, Ref. 1
$(1.00 \pm 0.02)\ P/(v/c)\ (Y^{90})$	double scattering	HEINTZE [1]
-0.981 ± 0.012	electrostatic deflection and Mott scattering	KETELLE [2]
-0.76 ± 0.15	electrostatic deflection and Mott scattering	DE WAARD [4]
-0.94 ± 0.06	Møller scattering	GEIGER, p. 59, Ref. 3
-1.08 ± 0.21	$B + E$ field, and Mott scattering	CAVANAGH, p. 55, Ref. 1
-0.98 ± 0.11	$B + E$ field, Mott scattering and preacceleration	CAVANAGH cf. Fig. 53
-0.82 ± 0.15	electrostatic deflection and Mott scattering	FRAUENFELDER, p. 53, Ref. 3
-0.96 ± 0.06	electrostatic deflection and Mott scattering	BIENLEIN, p. 54, Ref. 1
-0.74 ± 0.15	electrostatic deflection and Mott scattering	DE WAARD [4]
$(1.00 \pm 0.015)\ P/(v/c)\ (Co^{60})$	electrostatic deflection and Mott scattering	DE WAARD [5]
-0.93 ± 0.21	Møller scattering	BENCZER-KOLLER, p. 59, Ref. 2
-0.86 ± 0.06	Møller scattering	GEIGER, p. 59, Ref. 3
-0.99 ± 0.037	double scattering	ALIKHANOV cf. Fig. 56
-0.82 ± 0.15	double scattering	HEINTZE [1]
-1.02 ± 0.15 -1.15 ± 0.4	} $B + E$ field and Mott scattering	ALIKHANOV, p. 55, Ref. 2
high negative	bremsstrahlung and Compton scattering	SCHOPPER, p. 65, Ref. 2
high negative	bremsstrahlung and transmission	GOLDHABER, p. 65, Ref. 1
-0.97 ± 0.20	$B + E$ field and Mott scattering	CAVANAGH, p. 59, Ref. 3
same as P^{32}	electrostatic deflection and Mott scattering	DE WAARD, [5]
same as P^{32}	double scattering	DE SHALIT, p. 57, Ref. 1
-1.02 ± 0.19	Møller scattering	BENCZER-KOLLER, p. 59, Ref. 2
-0.98 ± 0.18	Møller scattering	GEIGER, p. 59, Ref. 3
$(0.97 \pm 0.04)\ P/(v/c)\ (Sr^{90} + Y^{90})$	double scattering	HEINTZE [1]
-1.0 ± 0.10	double scattering	ALIKHANOV cf. Fig. 56
$(0.99 \pm 0.03)\ P/(v/c)\ (Sr^{90} + Y^{90})$	double scattering	HEINTZE [1]
-0.53 ± 0.15	electrostatic deflection and Mott scattering	DE WAARD [4]
-1.0 ± 0.10	double scattering	ALIKHANOV cf. Fig. 56
-1.0 ± 0.10	double scattering	ALIKHANOV cf. Fig. 56
-1.0 ± 0.10	double scattering	ALIKHANOV cf. Fig. 56
$(1.016 \pm 0.014)\ P/(v/c)\ (Tl^{204})$	double scattering	HEINTZE [6]
$(0.99 \pm 0.02)\ P/(v/c)\ (P^{32})$	double scattering	BÜHRING [7]

Table 4

Decay	Character	E_{Max} kev	Electron E kev	$\langle v/c \rangle$
$RaE \xrightarrow{\beta^-} Po^{210}$ $(1^- \to 0^+)$	first forbidden	1160	200 to 600 >300 different	~0.8 >0.78 different
$Pr^{144} \xrightarrow{\beta} Nd^{144}$ $(0^- \to 0^+)$	first forbidden Fermi	3000	400 to 1100 1200 to 3000 >300 200 to 400	0.86 0.97 >0.78 0.8
$Na^{22} \xrightarrow{\beta^+} Ne^{22}$ $(3^+ \to 2^+)$	allowed, pure G-T	540	>200	0.7
$Cu^{64} \xrightarrow{\beta^+} Ni^{64}$ $(1^+ \to 0^+)$	allowed, pure G-T	660		
$N^{13} \xrightarrow{\beta^+} C^{13}$ $(\frac{1}{2}^- \to \frac{1}{2}^-)$	allowed, mixed	2210		
			>500	>0.8
$Ga^{66} \xrightarrow{\beta^+} Zn^{66}$ $(0^+ \to 0^+)$	allowed, pure Fermi	4144		~1
			>2000	~1
$Cl^{34} \xrightarrow{\beta^+} S^{34}$ $\left(\begin{smallmatrix} 0^+ \to 0^+ \\ 3^+ \to 2^+ \end{smallmatrix}\right)$	allowed	~5000	>2000	~1

converter. Thus it is necessary in order to calculate the absolute value of the polarization to integrate the whole process over the stopping of the electrons, as regards the energy and angular distribution of the bremsstrahlung. The experiment was performed using $Sr^{90} + Y^{90}$. Again the result indicates a high degree of polarization. The sign is negative.

ε) *Conclusion.* The discussion of the beautiful electron polarization measurements may be concluded with Table 4. In this table some results concerning electron and positron polarization measurements are collected. Besides the results of the experiments which have been discussed, the table includes results of some experiments which ought to be discussed too, but which are rather similar in technique to those which have been described already.

Most of the results are not very accurate. The conclusion to be drawn from the experiments is that electrons as well as positrons from β-decay are polarized to a degree of v/c to within 10 percent or better. There is no reason to believe as has been done that Fermi and Gamow-Teller transitions do not behave in the same manner. The directions of spin and momentum are parallel for positrons and antiparallel for electrons.

15. The β-decay of polarized nuclei. The first demonstration of the lack of parity conservation in β-decay was given by Wu et al.[1] in an experiment following a suggestion made by Lee and Yang. The idea underlying the experiment is that the *angular distribution of β-particles from polarized nuclei* will not be isotropic if parity is not conserved, whereas parity conservation requires an isotropic

---

[1] C.S. Wu, E. Ambler, R.W. Hayward, D.D. Hoppes and R.P. Hudson: Phys. Rev. **105**, 1413 (1957); **106**, 1361 (1957); **108**, 503 (1957).

(Continued).

$\dfrac{P}{v/c}$	Method	Referenc
$(0.83 \pm 0.02)\ P/(v/c)\ (\text{Tl}^{204})$	double scattering	HEINTZE [6]
-0.69 ± 0.10	Møller scattering	GEIGER, p. 59, Ref. 3
see Fig. 47	double scattering	ALIKHANOV, p. 48, Ref. 1
$\left.\begin{array}{l}-0.77 \pm 0.21 \\ -1.08 \pm 0.26\end{array}\right\}$	Møller scattering	FRAUENFELDER [3]
-0.90 ± 0.22	Møller scattering	GEIGER, p. 59, Ref. 3
$(0.96 \pm 0.04)\ P/(v/c)\ (\text{Sr}^{90} + \text{Y}^{90})$	double scattering	HEINTZE [1]
$(0.986 \pm 0.030)\ P/(v/c)\ (\text{Y}^{90})$	Møller scattering	MEHLHOP [8]
$> +0.4$	positronium formation and angular correlation	PAGE, p. 63, Ref. 1
high positive	annihilation-in-flight and angular correlation	HANNA, p. 62, Ref. 1
high positive	annihilation-in-flight and angular correlation	HANNA, p. 62, Ref. 1
$+0.93 \pm 0.20$	circular polarization of annihilation radiation	BOEHM, p. 64, Ref. 2
high positive	annihilation-in-flight in magnetized iron	FRANKEL, p. 61, Ref. 1
high positive	circular polarisation of annihilation radiation	DEUTSCH, p. 64, Ref. 1
high positive	circular polarisation of annihilation radiation	DEUTSCH, p. 64, Ref. 1

Underlining of results indicates that the v/c-dependence has been checked.

[1] J. HEINTZE: Z. Physik **150**, 134 (1958); **148**, 560 (1957).

[2] H. B. WILLARD, A. R. BROSI, A. GALONSKI and B. H. KETELLE: ORNL-2910, 7 (1960).

[3] H. FRAUENFELDER, A. O. HANSON, N. LEVINE, A. ROSSI and G. DE PASQUALI: Phys. Rev. **107**, 643 (1957).

[4] H. DE WAARD and O. J. POPPEMA: Physica, Haag **23**, 597 (1957).

[5] H. DE WAARD, O. J. POPPEMA and I. VAN KLINKEN: Proc. Reh. Conf. Nucl. Structure, 1958, p. 388.

[6] J. HEINTZE and W. BÜHRING: Z. Physik **153**, 237 (1958).

[7] W. BÜHRING: Z. Physik **155**, 566 (1959).

[8] W. A. W. MEHLHOP: OSR-TN-59-1323 (1960).

distribution, i.e. $A = 0$ in the distribution law, which has been shown to be

$$W(\vartheta) = 1 + A\,\frac{v}{c}\,\frac{\langle J_z \rangle}{J}\cos\vartheta. \qquad (15.1)$$

ϑ denotes the angle between the nuclear orientation direction and the particle emission direction. $\langle J_z \rangle$ is the expectation value of the spin component in the orientation direction. $\langle J_z \rangle / J$ is thus a measure of the degree of polarization of the initial nucleus. A is an important constant depending on the coupling constants and the relevant matrix elements

$$
\begin{aligned}
A = \Bigg\{ &\lambda_{JJ'}\Big[\pm \operatorname{Re}\left(C_T^* C_T' - C_A^* C_A'\right) - \frac{Z e^2}{\hbar c p}\operatorname{Im}\left(C_T^* C_A' + C_T'^* C_A\right)\Big]|\!\int\!\vec\sigma|^2 + \\
&+ \delta_{JJ'}\sqrt{\frac{J}{J+1}}\Big[\operatorname{Re}\left(C_T^* C_S' + C_T'^* C_S - C_A^* C_V' - C_A'^* C_V\right) \pm \\
&\pm \frac{Z e^2}{\hbar c p}\operatorname{Im}\left(C_A^* C_S' + C_A'^* C_S - C_T^* C_V' - C_T'^* C_V\right)\Big]|\!\int\!1|\,|\!\int\!\vec\sigma|\Bigg\}\frac{2}{\xi(1+\beta/W)}
\end{aligned}
\qquad (15.2)
$$

where

$$\lambda_{JJ'} = \begin{cases} 1 & J \to J' = J - 1 \\ \dfrac{1}{J+1} & J \to J' = J \\ -\dfrac{J}{J+1} & J \to J' = J + 1. \end{cases} \qquad (15.3)$$

The upper sign is used for β^--decay, the lower for β^+-decay. ξ and β are given by Eqs. (20.2) to (20.4).

In the experiment of Wu et al. the nuclear orientation is obtained by use of the Rose-Gorter me'hod. This method uses *adiabatic demagnetization*. A paramagnetic salt is suspended in the inner Dewar vessel of the cryostat (see Fig. 66) by means of thermo-insulating material. At the beginning of the cooling cycle, heat contact between the salt and the liquid helium surrounding the inner Dewar vessel is established by means of a helium atmosphere. A strong magnetic field is applied to the salt. Due to this field the electronic configuration of the salt is oriented. During the orientation process heat is developed in the salt and transferred to the surrounding liquid helium through the heat contact. When the orientation process is stopped and temperature equilibrium is obtained at liquid helium temperature 1.2° K, the helium atmosphere is removed by pumping, that is the heat contact is "switched off". Now the magnetic field is diminished and the salt is thus demagnetized adiabatically. The demagnetization causes a further lowering of the temperature of the salt. In this way temperatures below 0.01° K may be obtained. If the magnetic field is not completely removed, a considerable cooling is still obtained, but the electrons in the unfilled shells of the paramagnetic ions are still oriented and the hyperfine structure coupling will cause a polarization of the nuclei.

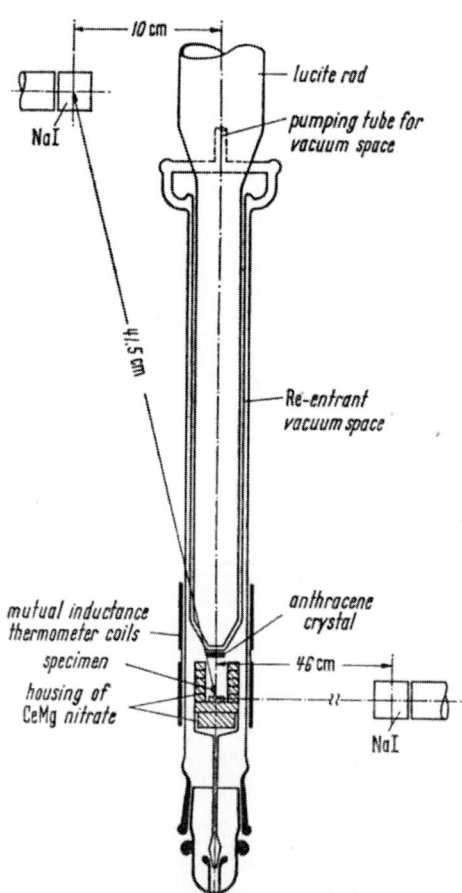

Fig. 66. The cryostat used by Wu et al. for the measurement of the angular distribution of β-particles from oriented nuclei. (After C.S. Wu, E. AMBLER, R.W. HAYWARD, D. HOPPES and R.P. HUDSON.)

The γ-ray emission from oriented Co⁶⁰ nuclei has been extensively studied. The emission is known to show a considerable asymmetry. This effect was used by Wu et al. in order to measure the degree of polarization. The γ-ray asymmetry was measured by means of two NaI(Tl) scintillation counters also shown in Fig. 66. The effect of the magnetic field on the gain of the photomultipliers was shown to be neglegible. When the degree of polarization is known, that is $\langle J_z \rangle / J$ is known, the β-decay asymmetry is a measure of the coupling constant combination describing the asymmetry through the constant A. The γ-emission asymmetry and the β-decay asymmetry are shown in Fig. 67 as functions of time. The result of the experiment is extracted from such curves. The result can be expressed by the statement that A [Eq. (15.1)] is approximately equal to -1, thus clearly demonstrating that parity is not conserved.

The β-decay of oriented nuclei has been studied on a series of nuclei notably those of isotopes of Co and Mn and the neutron. The neutron decay is discussed in Sect. 17[1].

The technique used in these experiments is rather difficult. Furthermore the applicability of the experiments involving nuclear orientation is restricted by the fact that only few materials can be oriented although new methods have been tried during the last few years. The accuracy of the type of experiments discussed here is not too good. Thus the main interest in these experiments for the time being is that of establishing the existence of the parity-non-conserving terms in the β-decay Hamiltonian. In order to evaluate the coupling constants one has to rely on other methods.

16. β-γ-polarization directional correlations. There is a close connection between the angular distribution of β-particles from oriented nuclei and the *β-γ-polarization directional correlation* for decays of unpolarized nuclei. The following remarks may clarify this situation. It is obvious that when the distribution of β-particles from polarized nuclei is asymmetric, that is the nuclear spin direction and the β-particle momentum are preferentially parallel or antiparallel, β-particles selected in a certain direction will leave the product nuclei polarized parallel or antiparallel to the selected direction respectively. The spin orientation will show a pattern like

$$1 + \frac{v}{c}\,\frac{\langle J_z \rangle}{J}\,A\cos\vartheta \tag{16.1}$$

where A is positive or negative respectively. This spin orientation pattern will influence a subsequent γ-ray. This sort of spin pattern will not influence the angular distribution of the γ-rays. But the circular polarization of γ-quantum in a certain direction will depend on the spin orientation pattern in the intermediate state. Thus a measurement of β-γ-circular polarization-directional

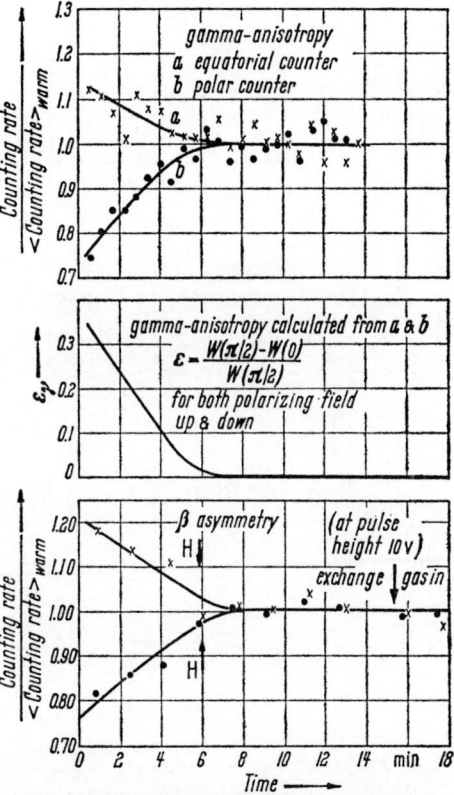

Fig. 67. The β- and γ-ray asymmetry as a function of time as measured by Wu et al. with the instrument shown in Fig. 66. (After C. S. Wu, E. Ambler, R. W. Hayward, D. D. Hoppes and R. P. Hudson.)

---

[1] Concerning the decay of oriented nuclei other than the neutron see also e.g. P. Dagley, M. A. Grace, J. S. Hill and C. V. Sowter: Phil. Mag., Ser. VIII, **3**, 489 (1958). — H. Postma, W. J. Huiskamp, A. R. Miedema, M. J. Steenland, H. A. Tolhoek and C. J. Gorter: Physica, Haag **23**, 259 (1957).

correlations gives the same type of information as that given by the angular distribution of β-particles from oriented nuclei. The information to be gained concerns the parity-non-conserving terms in the Hamiltonian, interference effects between Fermi and Gamow-Teller interactions, and information on the different nuclear matrix elements especially for first forbidden decays[1].

The angular correlation between the β-particles and circularly polarized γ-quanta will be

$$W(\vartheta) = 1 \pm \frac{v}{c} A \cos \vartheta \tag{16.2}$$

where the $+$ and $-$ signs are used for right and left circular polarization respectively.

The degree of circular polarization of γ-quanta emitted at an angle ϑ with respect to the β-particles will be

$$P = A \frac{v}{c} \cos \vartheta. \tag{16.3}$$

The A of these equations is given by

$$\left. \begin{aligned} A = \frac{1}{L+1} &\left\{ \mu_{JJ'} \left[\pm \mathrm{Re}\,(C_T^* C_T' - C_A^* C_A') - \frac{Z\,e^2}{\hbar\,c\,p} \mathrm{Im}\,(C_T^* C_A' + C_T'^* C_A) \right] \left| \int \vec{\sigma} \right|^2 + \right. \\ &+ \delta_{JJ'} \sqrt{\frac{J+1}{J}} \left[\mathrm{Re}\,(C_T^* C_S' + C_T'^* C_S - C_A^* C_V' - C_A'^* C_V) \pm \right. \\ &\left. \pm \frac{Z\,e^2}{\hbar\,c\,p} \mathrm{Im}\,(C_A^* C_S' + C_A'^* C_S - C_T^* C_V' - C_T'^* C_V) \right] |\int 1||\int \vec{\sigma}| \left. \right\} \frac{2}{\xi(1+\beta/W)} \end{aligned} \right\} \tag{16.4}$$

where

$$\mu_{JJ'} = \begin{cases} 1 & J \to J' = J - 1 \\ -\dfrac{1}{J} & J \to J' = J \\ -\dfrac{J+2}{J+1} & J \to J' = J + 1 \end{cases} \tag{16.5}$$

and ξ and β are again given by Eqs. (20.2) to (20.4).

The important experimental information to be obtained is the correct energy and angular dependence of the polarization and the value of A which depends on the β-decay coupling constants and the relevant matrix elements. A is not exactly identical with the constant A in the β-distribution from oriented nuclei (see Sect. 15). Of course the value of the information on the nuclear matrix elements which can be gained from a measurement of A depends on our knowledge of the features of β-decay and vice versa. Most β-γ-polarization experiments intends to gain information on the β-decay interaction. The determination of nuclear matrix elements will not be discussed here, but for the sake of completeness we may refer to a work by Steffen[2]. He measured the directional and the polarization-directional correlations in first forbidden transitions (Au[198]) in order to separate the different nuclear matrix elements.

The experimental problems need no long discussion. The γ-polarization can be measured by use of one of the γ-polarization measuring methods discussed in Sect. 14δ. The method which has been most extensively used is the forward

[1] T. Kotani and M. Ross: Phys. Rev. 113, 622 (1959).
[2] R.M. Steffen: Phys. Rev. 118, 763 (1960).

Compton scattering method (see Fig. 65)[1]. The transmission method has been used too[2].

STEFFEN[3] used another variant of the Compton scattering method. In the forward scattering experiments of p. 65, ref. 2 the electron-spin direction and

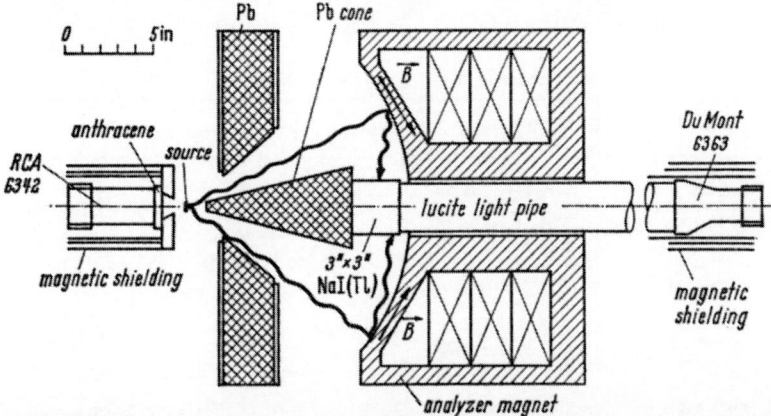

Fig. 68. The instrument used by STEFFEN for the measurement of circular polarization of γ-rays by Compton scattering in magnetized iron. The incoming γ-ray is perpendicular to the electron spin. (After R. M. STEFFEN.)

the photon momentum are almost parallel. In the Steffen instrument (Fig. 68) these directions are perpendicular. In this geometry the scattering rates of circularly polarized photons are different parallel and antiparallel to the electron spin direction. The optimum efficiency of this method is for a photon energy of about 400 kev against that of forward scattering which requires higher energy.

A common difficulty for γ-ray po-larisation measurement is that, due to the small differences to be measured, a very high stability of the counting ap-paratus is required. A procedure com-monly adopted in order to diminish the stability requirement is that of chang-ing the current direction of the polar-ization analyzer periodically with short time intervals, say 1 to 10 min. In this way the drifts may be partly balanced. Furthermore extensive shielding of the photomultiplier against the magnetic stray fields is necessary.

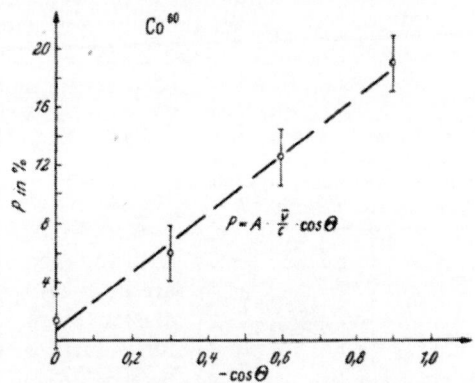

Fig. 69. The circular polarization of a γ-ray emitted in a direction forming the angle ϑ with the direction of a previously emitted β-ray, as a function of cos ϑ as measured by APPEL. (After H. APPEL.)

The results of the β-γ-polarization directional correlation measurements are not very accurate. Few of the measurements are better than 20% of the measured effects. Some of the experimental results are summarized in Table 5. This table is not intended to be complete, only illustrative.

The cos ϑ dependence of the polarization is quite well established. As an example Fig. 69 shows the result of a measurement of the angular dependence

[1] F. BOEHM and A. H. WAPSTRA: Phys. Rev. **106**, 1364 (1957). — H. SCHOPPER: Phil. Mag. **2**, 710 (1957). — TH. MAYER-KUCKUK and R. NIERHAUS: Z. Physik **154**, 383 (1959).
[2] A. LUNDBY, A. P. PATRO and J. P. STROOT: Nuovo Cim. **6**, 745 (1957). — L. A. PAGE, B.-G. PETTERSON and T. LINDQUIST: Phys. Rev. **112**, 893 (1958).
[3] See footnote 1, p. 52.

performed by Appel[1]. The v/c dependence has not been unambigously proved. The experiment made by Page, Patterson and Lindquist (p. 73, Ref. 2) showed at an early stage of the experiments that the correlation did not depend on v/c. Later measurements, however, seems to indicate the expected linear relation. As an example Table 6 shows the results of a measurement made by Lee, Benczer-Koller and Wu[2] on Co⁶⁰. The Fermi-Gamow-Teller interference term

Table 5. *Some results of β-γ-polarization correlation experiments.*

Decay	Measured A	Theoretical A assuming G-T	Reference	Remarks
Co⁶⁰ 5⁺ $\xrightarrow{\beta^-}$ 4⁺ → 2⁺ → 0⁺	− 0.32 ± 0.07 − 0.41 ± 0.07 − 0.40 ± 0.09 − 0.335 ± 0.018 − 0.37 ± 0.06	− 0.33	3 4 5 6 7	pure G-T β^--decay
Na²² 3⁺ $\xrightarrow{\beta^+}$ 2⁺ → 0⁺	+ 0.34 ± 0.14 + 0.39 ± 0.08 + 0.295 ± 0.054	+ 0.33	3 4 6	pure G-T β^+-decay
Sc⁴⁶ 4⁺ $\xrightarrow{\beta^-}$ 4⁺ → 2⁺ → 0⁺	+ 0.33 ± 0.04 + 0.23 ± 0.06 + 0.29 ± 0.11 + 0.24 ± 0.04	+ 0.08	5 7 3 8	{mixed Fermi and G-T. Shows interference
Au¹⁹⁸ 2⁻ $\xrightarrow{\beta^-}$ 2⁺ → 0⁺	+ 0.45 ± 0.06 + 0.42 ± 0.10		5 7	{first forbidden. Shows interference

[i.e. the term containing $|\int 1| \cdot |\int \vec{\sigma}|$ in Eq. (16.3)] seems to be maximum. Boehm and Wapstra[9] measured A for Sc⁴⁶ with the result $+ 0.33 \pm 0.04$. The assumption of no interference leads to $+ 0.08$. Further discussions on the results of the experiments as regards the coupling constant determination are postponed to Sect. 20.

Table 6. *v/c dependence of the ratio between experimental results on the Co⁶⁰ β-γ-polarization (Wu et al.[2]) and the theoretical expectation.*

v/c	P_{exp}/P_{theory}
0.40	1.14 ± 0.23
0.60	1.00 ± 0.20
0.72	1.03 ± 0.21

Note added in proof. Bloom, Mann and Miskel[10] and Daniel and Kuntze[11] have recently remeasured the A of Sc⁴⁶. Their results are $+ 0.087 \pm 0.017$ and $+ 0.10 \pm 0.02$, respectively. These values fit nicely with the theoretical A assuming $|\int 1| = 0$. These results are interesting not only because they make the conclusion of maximum Fermi-G-T interference doubtful. A further assumption needed in order to interpret the large asymmetry earlier found was that $|\int \vec{\sigma}| = 0$. This indicated a violation of the isotopic spin selection rule for the Fermi *matrix* element. This violation is not necessary with the new results in mind.

[1] H. Appel: Z. Physik 155, 580 (1959).
[2] Y. K. Lee, N. Benczer-Koller and C. S. Wu: Bull. Amer. Phys. Soc. II 5, 9, (1960).
[3] A. Lundby, A. P. Patro and J. P. Stroot: Nuovo Cim. 6, 745 (1957); 7, 891 (1958).
[4] H. Schopper: Phil. Mag., Ser. VIII 2, 710 (1957).
[5] F. Boehm and A. H. Wapstra: Phys. Rev. 106, 1364 (1957); 107, 1202 (1957). — Z. Physik 152, 384 (1958).
[6] H. Appel, H. Schopper and S. D. Bloom: Phys. Rev. 109, 2211 (1958).
[7] R. M. Steffen: Proc. Rehovoth Conf. Nucl. Structure p. 419. Amsterdam: North-Holland Publishing Co. 1958.
[8] W. Jüngst and H. Schopper: Z. Naturforsch. 13a, 505 (1958).
[9] F. Boehm and A. H. Wapstra: Phys. Rev. 109, 457 (1958).
[10] S. D. Bloom, L. G. Mann and J. A. Miskel: Phys. Rev. Letters 5, 326 (1960).
[11] H. Daniel and M. Kuntze: Z. Physik 162, 229 (1961).

17. The neutron decay. Since the theory of β-decay is based on the idea of neutrons decaying into protons, electrons and neutrinos, a considerable support for the theory was obtained when the β-decay of the free neutron was observed by SNELL and MILLER and by ROBSON[1].

The experiments leading to this discovery represent a considerable effort. Neutrons from a nuclear reactor will decay during their flight so that the number of decays per cm of the neutron beam will be inversely proportional to the velocity of the neutrons and proportional to the intensity of the beam. Even with slow neutrons it is seen that a considerable beam intensity is needed. The necessary intensity information is conveniently measured by means of a $1/v$ capture cross section for neutrons. However, the large amount of γ-rays and secondary electrons, present in the neighborhood of a reactor and a neutron beam, gives rise to a considerable background against which the decay electrons or recoils must be measured. The first experiments carried out by SNELL and MILLER were the observation of low energy positively charged ions originating from the neutron beam in vacuum. The later experiments were carried out by observing the proton recoils in coincidence with the electrons and identifying the recoils as protons. In the work of ROBSON this was done by means of a proton spectrometer.

The neutron decay not only represents a support to the β-theory. In itself it is perhaps the most important nuclear β-decay. There are two reasons. The neutron which is a simple particle decays by emission of a β^--particle and an (anti-)neutrino to a proton which is an equally simple particle. In any other β-decaying nuclei there are more neutrons and protons being able to transform into each other in a β-decay. The effect of this is that any term depending on the nuclear states, i.e. nuclear matrix elements, can be easily calculated with some precision in the neutron decay whereas the matrix elements of more complicated nuclei can usually be calculated in rough approximation only. Besides this fact more types of experiments yielding valuable information are possible in the neutron case than in case of any other nucleus. This remark applies to the experiments on the *decay of polarized nuclei* which involves detection of the recoil nucleus. These experiments are for the moment only possible in the neutron case. In principle all coupling constants can be evaluated from neutron experiments.

Because of these facts the neutron experiments will be discussed in rather great detail. In order to understand the different experiments which are going to be discussed, consider the decay probability for a neutron in a beam whose polarization is determined by $\langle \vec{J} \rangle / J$ decaying by emission of an electron with energy W in the solid angle $d\Omega_\beta$ around the direction Ω_β and a neutrino with energy $W_\nu = W_0 - W$ in the solid angle $d\Omega_\nu$ around the angle Ω_ν where W_0 is the total decay energy available. The probability of this process is given by:

$$P(\langle \vec{J} \rangle | W, \vec{p}_\beta, \vec{p}_\nu)\, dW\, d\Omega_\beta\, d\Omega_\nu = \frac{1}{(2\pi)^5}\, p_\beta\, W(W_0 - W)^2\, dW\, d\Omega_\beta\, d\Omega_\nu \times$$

$$\times \xi \left\{ 1 + \alpha\, \frac{\vec{p}_\beta \cdot \vec{p}_\nu}{W(W_0 - W)} + \beta\, \frac{1}{W} + \frac{\langle \vec{J} \rangle}{J} \cdot \left[A\, \frac{\vec{p}}{W} + B\, \frac{\vec{p}_\nu}{W_0 - W} + D\, \frac{\vec{p}_\beta \times \vec{p}_\nu}{W(W_0 - W)} \right] \right\}. \tag{17.1}$$

The constants $\xi, \alpha, \beta, A, B, D$ are different combinations of coupling constants and matrix elements [see Eqs. (20.2) to (20.4); (6.5); (21.1) to (21.3)].

[1] A. H. SNELL and L. C. MILLER: Phys. Rev. **74**, 1217 (1948). — A. H. SNELL, F. PLEASONTON and R. V. McCORD: Phys. Rev. **78**, 310 (1950). — J. M. ROBSON: Phys. Rev. **78**, 311 (1950).

The idea of the experiments is to measure these constants separately. The constants are singled out by suitable choices of combinations of the $\vec{p}$'s and $\langle \vec{J} \rangle / J$. The constants ξ, α, β are the classical constants, which are connected to the usual ft-value, electron-neutrino correlation, and Fierz interference term determinations and have no bearing on polarization or parity problems. If

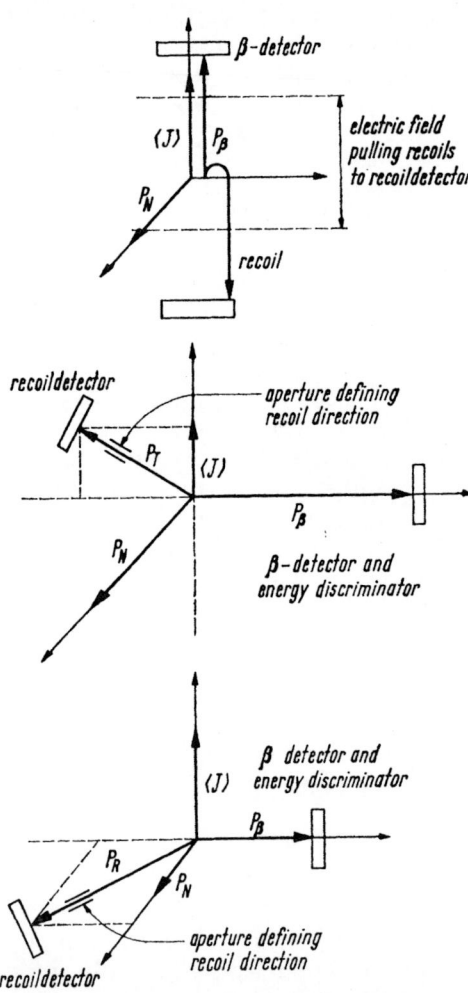

$\langle \vec{J} \rangle = 0$ (unpolarized beam) and only the total number of β-particles or recoil nuclei are measured, the ft value or the ξ can be determined. This measurement consists in principle of a measurement of the specific activity of a "sample" of known geometry and neutron density[1]. The value of the Fierz-interference term could in principle be determined from the form of the electron energy spectrum. This has not been done very exactly, but evidence from many other β-decays (see Sects. 12 and 20) seems to indicate that $\beta=0$ with reasonable certainty. The constant α is determined from a familiar recoil experiment[2].

If the neutron beam is polarized (i.e. $\langle \vec{J} \rangle \neq 0$) A, B and D can be measured. The principle of the experiment measuring A is shown in Fig. 70a. $\dfrac{\langle \vec{J} \rangle}{J} \cdot \vec{p}_\beta$ is maximum and all terms containing $\vec{p}_\nu$ average to zero because all recoil directions are recorded. The measurement of B is illustrated in Fig. 70b. In this arrangement the β-particle direction and energy is measured. The direction of the recoil, but not the energy is singled out by means of a diaphragm. Thus the momentum direction and magnitude of all decay particles are uniquely determined by energy and momentum conservation. Conditions are chosen so that $\vec{p}_\nu$

Fig. 70a–c. Principles of experiments on the decay of polarized neutrons. a Neutron spin—electron correlation (A). b Neutron spin—neutrino correlation (B). c Neutron spin—$\vec{p}_\beta \times \vec{p}_\nu$ correlation (D).

is parallel or antiparallel to $\langle \vec{J} \rangle$. Here $\vec{p}_\beta \cdot \dfrac{\langle \vec{J} \rangle}{J}$ and $(\vec{p}_\nu \times \vec{p}_\beta) \cdot \dfrac{\langle \vec{J} \rangle}{J}$ equal zero. $\vec{p}_\nu \cdot \dfrac{\langle \vec{J} \rangle}{J}$ is maximum. The measurement of D is illustrated in Fig. 70c. $\vec{p}_\nu$ is defined as in the B-measurement. $\vec{p}_\beta \cdot \dfrac{\langle \vec{J} \rangle}{J} = 0, \vec{p}_\nu \cdot \dfrac{\langle \vec{J} \rangle}{J} = 0$, and $(\vec{p}_\beta \times \vec{p}_\nu) \cdot \dfrac{\langle \vec{J} \rangle}{J}$ is maximum.

[1] J.M. ROBSON: Phys. Rev. 83, 349 (1951). — A.N. SOSNOVSKY, P.E. SPIVAK, Y.A. PROKOFIEV, I.E. KUTIKOV and Y.P. DOBRININ: Nuclear Phys. 10, 395 (1959).
[2] J.M. ROBSON: Phys. Rev. 100, 933 (1955). — Canad. J. Phys. 36, 1450 (1958). — YU.V. TREBUKHOVSKIĬ, V.V. VLADIMIRSKIĬ, V.K. GRIGOR'EV and V.A. ERGAKOV: J. Exp. Theor. Phys. USSR. 36, 1314 (1959). — JETP [translation] 9, 931 (1959).

Thus it is seen that all the constants can be measured. With these illustrations in mind we may now turn to the discussion of the experimental arrangements.

α) *The neutron half-life.* The half-life (i.e. the ξ) has been measured by ROBSON and by SOSNOVSKI et al. (p. 16, Ref. 1).

ROBSON measured the energy spectrum of the β-decay of the neutron and the neutron half-life. His instrument is illustrated in Fig. 71. The instrument essentially consists of two magnetic spectrometers arranged in coincidence around a neutron beam. By means of an electric field all the protons from the source volume are extracted with nearly the same energy. Subsequently these recoil protons are sent into a recoil spectrometer to be focused on the recoil detector

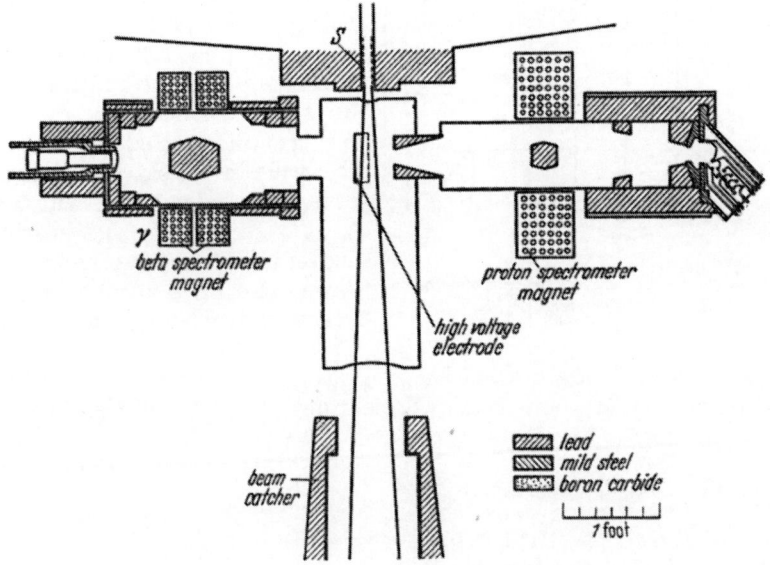

Fig. 71. ROBSON's arrangement for the measurement of the β-spectrum of the neutron. (After J. M. ROBSON.)

which is a multiplier arrangement of the usual type. The spectrometer ascertain that the particles detected are really protons. The socond spectrometer is an electron spectrometer furnished with an anthracene-scintillation counter. The number of neutrons decaying per cm³ was determined in two ways. The proton detector count rate and the coincidence count rate are divided by the appropriate effective source volumes. These are determined by a study of the possible proton paths as found in a mechanical model equivalent to the proton collecting system. The volumes are different because of the proton-electron correlation, which tends to increase the coincidence volume (when an electron is registered, the predominant proton direction will be against the proton spectrometer). These two determinations give an internal check on the source volume determination. The neutron decay rate per cm³ and the neutron density determine the half-life. The neutron density was measured by means of Mn-foils. Mn is a $1/v$-absorber at neutron energies below 1 ev [see Eq. (17.4)]. The density distribution across the beam was measured relative to the density in the center. This distribution was used as a weight function when calculating the effective volumes. The absolute neutron density was found by calibrating the foils against standard Mn-foils which in turn were standardized in a known thermal beam. This standard beam was measured accurately with a small boron trifluoride chamber.

The Fermi plot obtained by ROBSON is shown in Fig. 72. The maximum energy derived from this curve is (782 ± 13) kev in excellent agreement with the

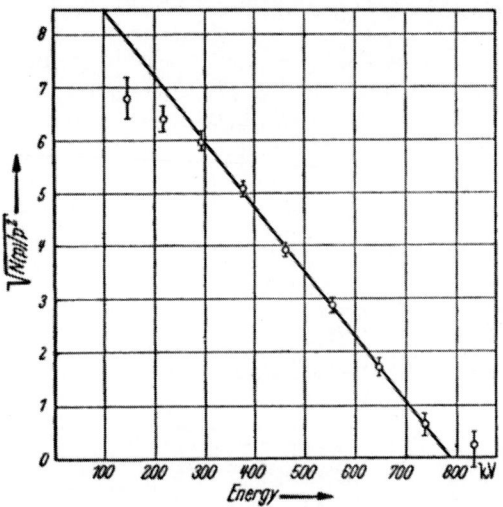

$n - p$ mass difference (782.3 ± 1.0) kev as found from reaction data. The half-life was found to be (12.8 ± 2.5) min.

Later SOSNOVSKY et al. (p. 16, Ref. 1) have remeasured the neutron half-life. Their instrument is shown in Fig. 73. A rotationally symmetric hollow high voltage electrode maintained at a high positive voltage (20 kev) is arranged coaxially around a well collimated neutron beam from a reactor. Protons created by a neutron decay in the central region of this electrode travel against the walls in a field free region. A tube, electrically ending with a grid 3, is attached perpendicularly to the central electrode at the opening 2. A proton created at point X in the hollow electrode reaches grid 3 with a probability Ω. Ω is the solid angle

Fig. 72 The Fermi plot for the neutron β-spectrum obtained by ROBSON using the instrument shown in Fig. 71. (After J. M. ROBSON.)

defined by that part of the grid 3 which is seen from point X through the aperture 2. The neutron density at point X is $n(X)$ neutrons per cm³, and the number of

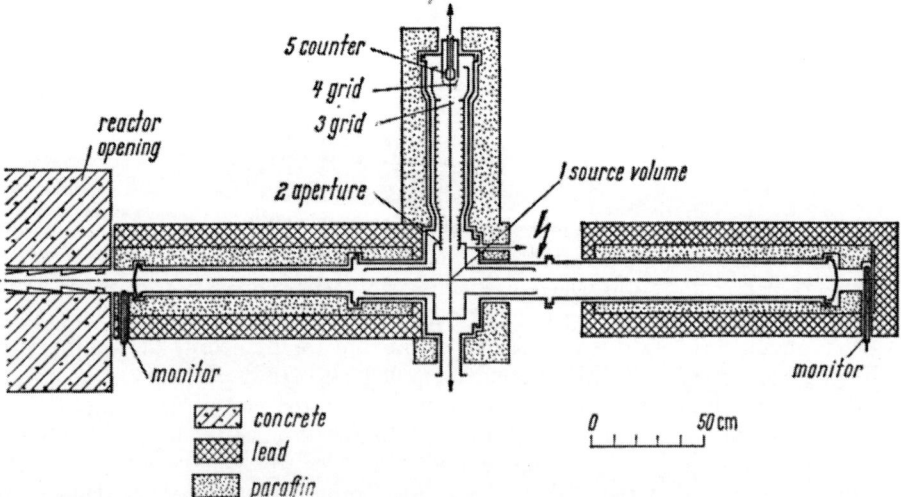

Fig. 73. The instrument used by SOSNOVSKY et al. for the measurement of the neutron half-life. (After A.N. SOSNOVSKY, P.E. SPIVAK, Y.A. PROKOFIEV, J.E. KUTIKOV and Y.P. DOBRONIN.)

decays per sec per cm³ is $\lambda \cdot n(X)$, where λ is the decay constant to be measured. Thus the total number of protons hitting grid 3 is

$$N = \int n(x, y, z) \, \lambda \Omega(x, y, z) \, dx \, dy \, dz. \tag{17.2}$$

Protons leaving grid 3 enter a strong field formed between grid 3 (20 kv) and a grounded semispherical grid 4. This field accelerates the protons, which are also focused by the same field onto the opening of the counter, a proportional counter placed behind grid 4. The counter opening is covered by a collodium film (0.07 μ) supported by a tungsten grid 5. In the counter a grid screens the active volume against charges accumulated by the window. The efficiency of the detector itself is 100%, but the total efficiency of the detector arrangement including grids is lower than that. This is due to the fact that protons may hit the grid wires and thus be lost. The total efficiency was estimated by calculation. The authors neglected the loss in grid 3, which was estimated to be less than one percent. The total efficiency was taken as the transparency of the remaining four grids. This transparency was again found as the product of the geometrical transparencies of the three grids. The calculated total efficiency was 0.843 ±0.06 obtained as the mean of 0.848 for the protons passing normal to the grids and 0.838 for the most obliquely incident protons. When this efficiency, η, is known, the N of Eq. (17.2) may be equated to n_p/η, where n_p is the number of protons which are recorded. From this the decay constant is calculated, when the neutron density has been measured. The calculation of the efficiency might seem questionable. The geometrical transparency of a grid is the part of the grid area which is not covered by wires. But the real transparency for a charged particle may well be different from the geometrical one owing to the fact that the electrical field is not homogeneous in the vicinity of the wire. These inhomogeneities may distort the trajectory of the particles. Furthermore a mere multiplication of the transparencies is a questionable procedure. There might easily be some coupling between the grids. This is to be understood so that a particle traversing the first grid may possibly be forced through the next grid too by the inhomogeneities of the field produced by all the grids. A careful measurement of the efficiency would be an improvement of the experiment.

Until now the connection between the measured proton rate and the neutron density has been discussed. The neutron density remains to be discussed. The spatial density of neutrons has to be known in order to evaluate the integral in Eq. (17.2). The distribution along the beam can be supposed to be constant, whereas that across the beam has to be measured. All neutron measurements were performed by activating Au and Na foils and measuring the induced activity. Actually the flux rather than the density determines the induced activity. This activity N_i per cm³ after irradiating a foil in t sec is:

$$N_i = \int n(v) \cdot v \cdot \Sigma(v)\, dv = \int \varphi(v)\, \Sigma(v)\, dv \qquad (17.3)$$

but when using a material with a $1/v$ activation cross section the induced activity will be:

$$N_i = \int n(v)\, v \cdot \Sigma_0 \cdot \frac{1}{v}\, dv = \Sigma_0 \int n(v)\, dv . \qquad (17.4)$$

Consequently the neutron density is measured. The density distribution across the beam was measured relatively by activating a gold foil covering the beam. Afterwards the foil was cut to pieces and the activity of each piece was measured. The density of the central section of the beam was then absolutely measured by means of gold and sodium foils. The Na-foil was calibrated against the Au-foil, the cross section of which is well known. The value of the Au-cross section was taken as $(98 \pm 1.5) \times 10^{-24}$ cm². The intercalibration was performed in a thermal beam, which was shown to be thermal by means of cadmium-ratio measurements (Na: 4000; Au: 150). The neutron density measured by Au respectively Na foils were consistent within ±0.5%. The authors claim the limits of error in the neutron density measurements to be ±1.8%. The neutron half-life as measured by this experiment was (11.7 ± 0.3) min. The quoted error includes statistics, accuracy of the neutron density measurement and the estimated error of the detector efficiency. The accuracy of this experiment is claimed to be better than that of Robson's experiment by an order of magnitude. This is a surprisingly large improvement in accuracy, but although we have permitted ourselves to doubt the grid efficiency calculations it should be stated that the experiment has been carried out with much care and we shall adopt the result $T_{\frac{1}{2}} = (11.7 \pm 0.3)$ min and $ft = 1187 \pm 35$ sec for later discussions of the coupling constants.

β) *The neutron recoil experiments.* The next experiment to be discussed is the electron-neutrino correlation experiment carried out by Robson. The instru-

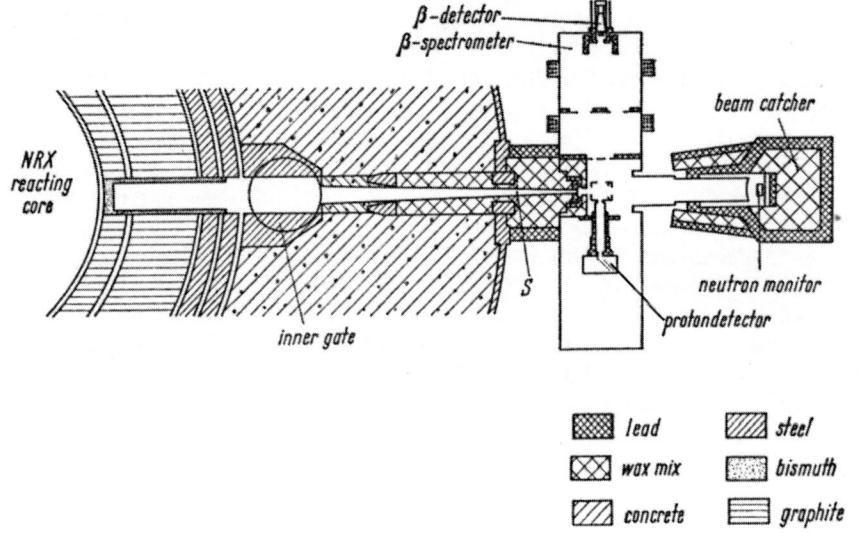

Fig. 74. An overall view of the arrangement used by Robson for the measurement of the β-ν correlation of the neutron decay. (After J.M. Robson.)

ment used by Robson is illustrated in Fig. 74. The proton counting part is illustrated in Fig. 75. The energy distribution of electrons from decays characterized by a certain angle between the electron and the recoil proton is measured. In Robson's experiment the direction of the β-particles is defined by two ring-shaped apertures A and B. The aperture placed between the β-spectrometer coils has only one essential function, that of minimizing all sorts of scattered radiation, both electrons and γ-rays. The β-spectrometer detector used is an anthracene scintillation counter feeding a single channel pulse height selector. The window of the selector was made as narrow as possible in order to reduce background from scattered radiation without loosing efficiency. The active source volume of the neutron beam is defined by the β-spectrometer. This source volume and the statistical weight at each point was measured by means of a movable radioactive source. The proton counter which is placed coaxially with the β-spectrometer is an electron multiplier of the Allen type. The protons created in the source volume travel against the wall in the field free interior of the centre electrode. Some of the protons have the correct direction allowing them to emerge through the gridded aperture which is placed in the wall opposite to the β-particle aperture. A positive potential (7 kv) is applied to the center electrode assembly. After travelling through a second grid covering the end of a tube section, which is also passed, the protons enter a second tube in electrical contact with the entrance aperture of the electron multiplier. The potentials of the center electrode, the intermediate tube and grid, and the second tube are adjusted so as to meet the following requirements. The protons are accelerated between the two grids through 1 kv. All protons created in the source volume and directed through the

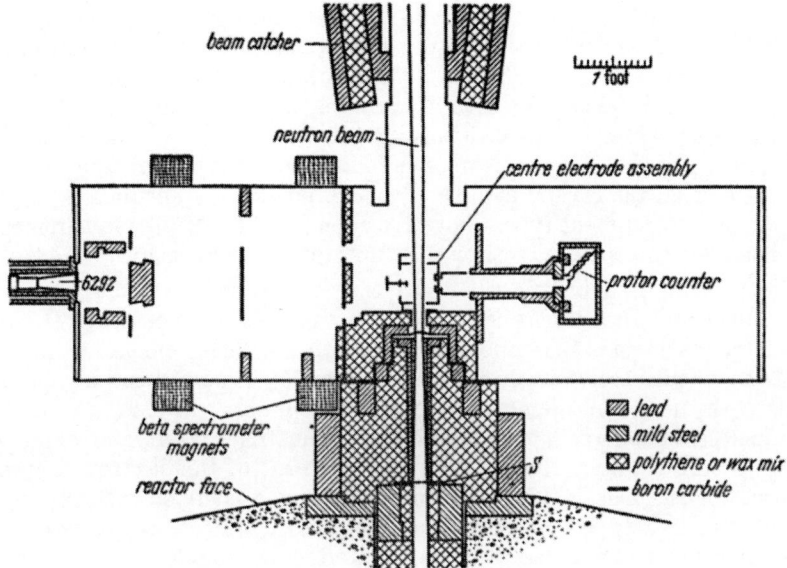

Fig. 75. Detector part of the instrument of Fig. 14. (After J. M. ROBSON.)

first grid reaches the gap between the two tubes without striking the walls. In this gab the protons are further accelerated through 6 kv and focused onto the entrance aperture. The focusing conditions were found by calculation. The potential of the entrance aperture was kept much above that of the cathode of the multiplier in order to stop secondary electrons created by stray radiation in the entrance aperture. Coincidences between β- and proton-detector were counted with various settings of the β-spectrometer. Every measuring point was found by extrapolating log (counting rate) against proton counter bias to zero bias. This curve was shown by experiment to be strictly linear. This method eliminated errors due to proton detector gain drift.

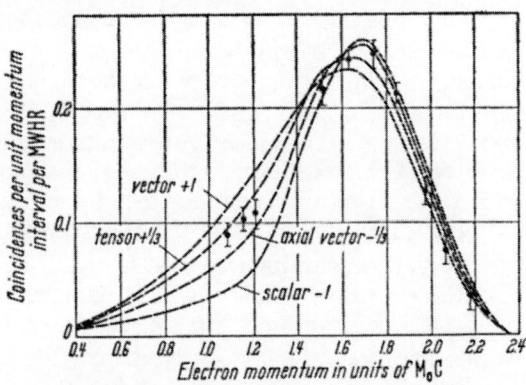

Fig. 76. The electron energy distribution measured by ROBSON by means of the instrument shown in Fig. 75. The curves are theoretical curves calculated for different values of α. (After J.M. ROBSON.)

Different monitors monitored the neutron beam. The neutron density distribution across the neutron beam was measured by activation of copper wires.

The β-energy spectrum to be expected from the apparatus was found by Monte Carlo calculations for different values of the correlation constant α. These curves were compared to the experimental points and the best fit of α was found. The value obtained was $+0.07 \pm 0.12$. The quoted error is a result of a careful analysis of all sources of error. Fig. 76 shows experimental points of one of the measurements and theoretical curves for the pure interactions. (p. 76, Ref. 2)

γ) *The decay of polarized neutrons.* The neutron experiments hitherto discussed are based on the use of an unpolarized neutron beam. The constants A, B and D found in Eq. (17.1) as coefficients to the terms containing the expectation value $\langle J \rangle$ of the spin of the decaying nuclei have all been measured in experiments based on the use of polarized neutron beams. Burgy et al. have measured both A, B and D[1] at Argonne. Clark and Robson have measured B[2] and recently also A[3] and Clark, Robson and Nathans have measured D[4]. In the discussion of these experiments it is found most convenient to group the measurements of all the constants made by one group of experimentalists instead of grouping after experiment type. Concerning the principle of the measurements the reader is referred to the discussion in connection with Figs. 70.

The instrument used by the Chalk River group is shown in Fig. 77. The specific arrangement shown here is that which was used for measuring D. The proton detecting system is the same as that used in the measurement of the electron neutrino angular correlation coefficient (Fig. 75). The β-spectrometer has been replaced by two 5 in. diameter by $\frac{1}{8}$ in. thick plastic scintillation counters. These are placed in a horizontal plane and in symmetrical positions around the axis of the proton counter. The angular displacement of the electron counters with respect to this axis is 160°. By means of single channel analyzers only electrons of energy from 350 to 550 kev are detected. Coincidences between the proton counter and either of the electron counters were recorded. The neutrons are polarized by transmission through magnetized iron. The polarization of the neutrons in the source volume was measured by diffracting the neutrons of wavelength 2.8 Å twice, from the (100) planes of a cooled Haematite antiferromagnetic crystal and from the (111) planes of a magnetized face centered cubic crystal of cobalt iron alloy. The average polarization was found to be 27% and the direction was up within 5° from the vertical. It is seen that $\langle \bar{J} \rangle \cdot \bar{p}_\beta = 0$ and $\langle \bar{J} \rangle \cdot \bar{p}_\nu = 0$ and consequently $\langle \bar{J} \rangle \cdot \bar{p}_\nu = \langle \bar{J} \rangle \cdot (-\bar{p}_\beta - \bar{p}_\nu) = 0$. For the proton-electron angle and the electron energy which was used, $\bar{p}_\beta \times \bar{p}_\nu$ is parallel to or antiparallel to the neutron spin for events involving electrons detected by the left or right β-detector respectively. In these cases $\langle \bar{J} \rangle \cdot (\bar{p}_\beta \times \bar{p}_\nu)$ is numerically maximum, but with opposite signs. Consequently the difference in coincidence rates for the two detectors is proportional to D. The instrumental asymmetry was measured with unpolarized neutrons obtained by switching off the magnetization of the polarizer. The final result of the D-experiment was $D = -0.14 \pm 0.20$.

The same apparatus was used in the B-measurement. The only change was that the electron counter assembly as a whole was rotated through 90° about the proton detector axes. In this case it is seen that $\bar{p}_\nu$ and $\bar{p}_\beta \times \bar{p}_\nu$ is rotated in the same way. Consequently the conditions are changed so that $\langle \bar{J} \rangle \cdot (\bar{p}_\beta \times \bar{p}_\nu) = 0$ while $\langle \bar{J} \rangle \cdot \bar{p}_\nu$ is maximum. In this case $\langle \bar{J} \rangle \cdot \bar{p}_\beta$ is not zero but small. A correction for the contribution of this term was applied to the result. The result of the experiment was $B = +0.96 \pm 0.40$. Recently Clark and Robson[3] have completed their neutron decay measurements by measuring A with the result -0.09 ± 0.05.

[1] M. T. Burgy, V. E. Krohn, T. B. Novey, G. R. Ringo and V. L. Telegdi: Phys. Rev. 120, 1829 (1960).
[2] M. A. Clark and J. M. Robson: Canad. J. Phys. 38, 693 (1960).
[3] M. A. Clark and J. M. Robson: Canad. J. Phys. 39, 13 (1961).
[4] M. A. Clark, J. M. Robson and R. Nathans: Phys. Rev. Letters 1, 100 (1958).

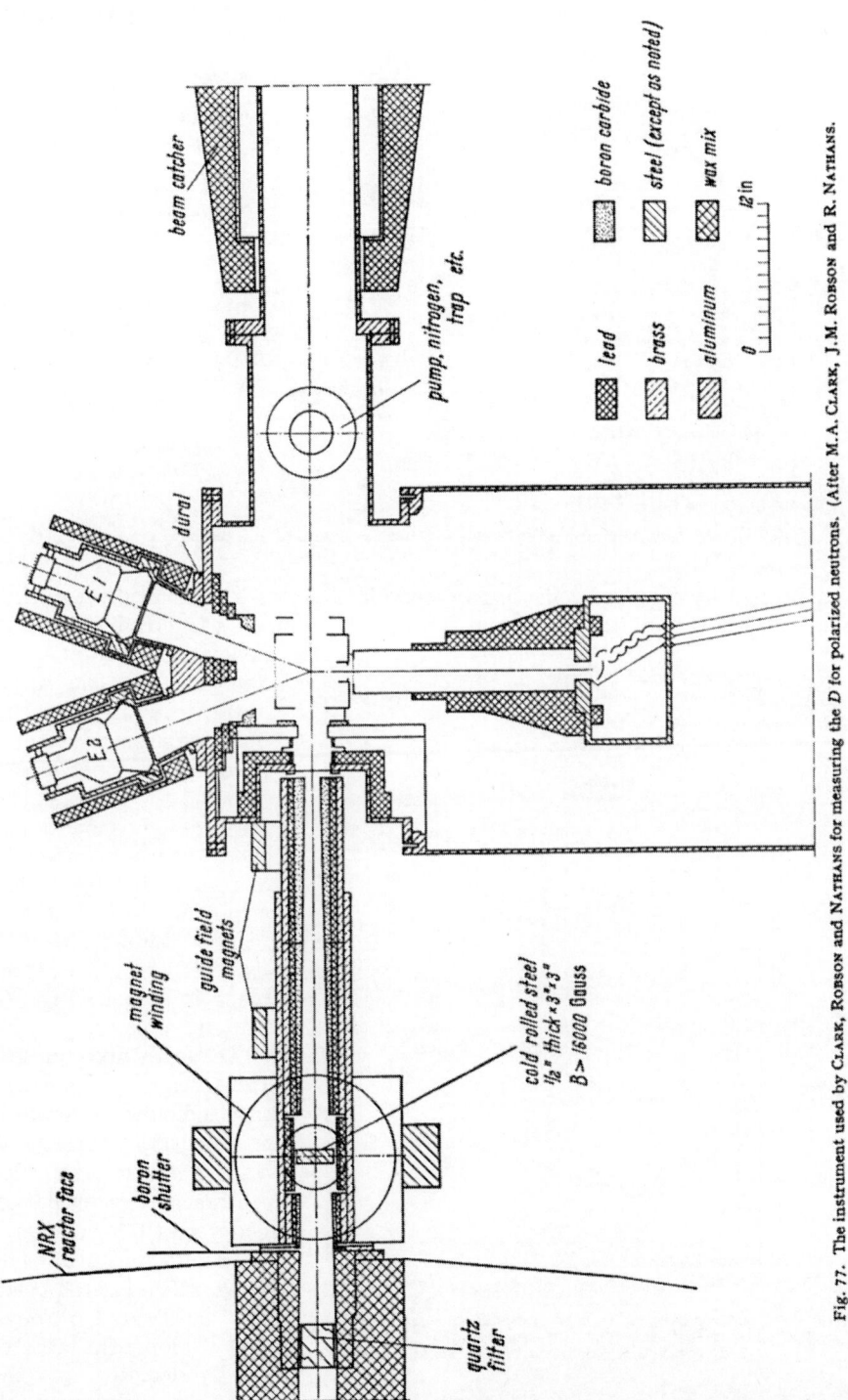

Fig. 77. The instrument used by Clark, Robson and Nathans for measuring the D for polarized neutrons. (After M.A. Clark, J.M. Robson and R. Nathans.

The Argonne group have measured all three constants A, B and D. The general physical layout of their instrument is seen in Fig. 78. The special variant

6*

of the instrument which is seen here is that used for the B and D measurement. The neutron beam is polarized by reflection from a cobalt mirror. The polarization

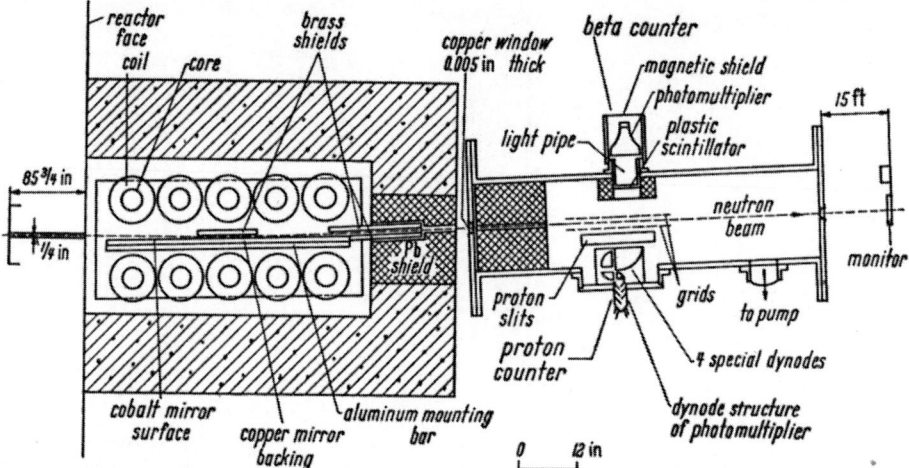

Fig. 78. Overall view of the instrument used by BURGY et al. for the experiments on the decay of polarized neutrons. (After M.T. BURGY, V.E. KROHN, T.B. NOVEY, G.R. RINGO and V.L. TELEGDI.)

was measured by a second reflection in a similar mirror. The counting rate after the double reflection was compared with that obtained when the beam was depolarized between the two mirrors. The depolarization was obtained by inserting a thin unmagnetized steel sheet in the beam between the mirrors. The measurement showed that the beam was polarized to a degree of (87 ± 7) percent. A cross sectional view of the detector part used for the B measurement is seen in Fig. 79. The neutrons are polarized vertically and the electrons leaving the beam perpendicular to the beam in the horizontal plane are detected by an Anthracene scintillation counter of 5 in. diameter by 0.2 in. thickness. The protons leaving the beam are accelerated through 800 V along the axis of the β-detector in the direction opposite to the β-detector. After traversing a grid the protons, now travelling along a field free path, reach a system of parallel slits allowing only protons having a certain downward component of momentum to

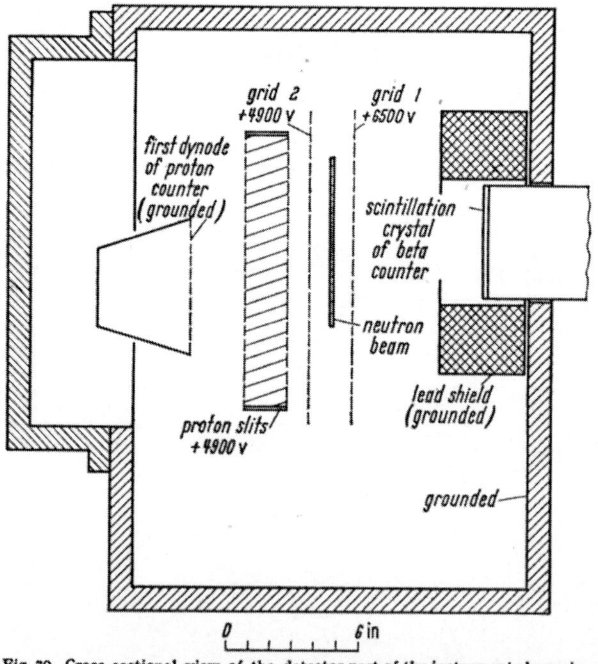

Fig. 79. Cross sectional view of the detector part of the instrument shown in Fig. 78. The B measuring version. (After M.T. BURGY, V.E. KROHN, T.B. NOVEY, G.R. RINGO and V.L. TELEGDI.)

pass. A proton which leaves this slit system is again accelerated through 5 kv and hits the cathode of an electron multiplier. It is seen that $\langle \bar{J} \rangle \cdot \bar{p}_\beta = 0$ and $\langle \bar{J} \rangle \cdot (\bar{p}_\beta \times \bar{p}_\nu) = \langle \bar{J} \rangle \cdot [\bar{p}_\beta \times (-\bar{p}_\beta - \bar{p}_r)] = 0$ in this geometry. Thus the coefficient B of the neutrino momentum term is really measured. Coincidences between the recoil detector and the β-detector are recorded for polarization up and down and for a beam which is depolarized by insertion of a steel sheet. The unpolarized beam experiment was made in order to facilitate the evaluation of the influence on the detectors of the magnetic field from the mirror. Of course the experimental conditions necessitated a careful calculation of the effect of finite solid angles, finite electron energy interval, dif-

ferent decay positions and change in proton momentum direction due to the acceleration. The result of this experiment was $B = 0.88 \pm 0.15$.

As in the Robson experiments discussed above a simple change in the B-apparatus made the D-measurement possible. The slits defining the proton momentum direction were rotated through $90°$ about the β-detector axis. In this case $\langle \bar{J} \rangle \cdot \bar{p}_\beta = 0$, $\langle \bar{J} \rangle \cdot \bar{p}_r = 0$ and consequently $\langle \bar{J} \rangle \cdot \bar{p}_\nu = 0$. The calculations in this case are very similar to those in the B-case, except that here the effect of the B-term has to be considered too. In the B-case the D-term gave no contribution because of the D being nearly zero. The actual D-value found was 0.04 ± 0.05.

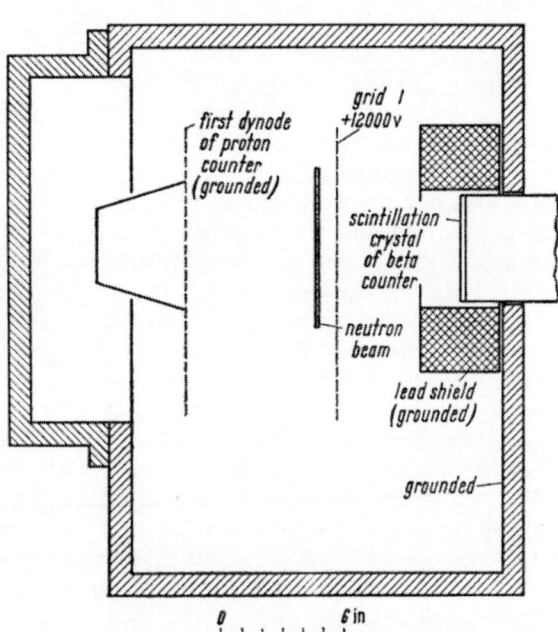

Fig. 80. Cross sectional view of the detector part of the instrument shown in Fig. 78. The A measuring version. (fter M.T. Burgy, V.E. Krohn, T.B. Novey, G.R. Ringo a .d N.L. Telegdi.)

The instrument used for the A-measurement was rather different (cf. Fig. 70). The experimental arrangement is shown in Fig. 80. The β-counter arrangement is the same as in the B and D-experiments. The neutron spin is changed by means of a magnetic guide field to left-right instead of up-down. As discussed in connection with Fig. 70 the B and D-terms in this case are not eliminated by choosing right angles between $\langle J \rangle$ and the appropriate momentum directions, but by integrating over all momentum directions but p_β. This is obtained by catching all recoils in the proton counter. To accomplish this the proton slits are removed and an accelerating field is applied between the neutron beam and the proton detectors. It is seen that $\langle \bar{J} \rangle \cdot \bar{p}_\beta$ is maximum and A is proportional to the counting rate difference for left and right polarization directions. The result had to be corrected for the B-term because the proton collection is not exactly 100%. The efficiency of the proton detector depends on the proton momentum. The result was $A = -0.114 \pm 0.019$.

Thus all the coefficients of Eq. (17.1) has been measured except β which is believed to be zero from other experiments (see Sect. 20). For the discussion of the coupling constants we adopt the following values for the different experimental results.

$$T_{\frac{1}{2}} = 11.7 \pm 0.3 \text{ sec}; \quad ft = 1187 \pm 35$$
$$a = 0.07 \pm 0.12$$
$$A = -0.114 \pm 0.019 \hspace{4em} (17.5)$$
$$B = 0.88 \pm 0.15$$
$$D = 0.04 \pm 0.05.$$

18. The neutrino helicity. The results of the electron polarization experiments and the determination of the coupling constants may be used for a determination of the neutrino helicity, i.e. the sign of the full polarization of the neutrino. The argument runs as follows. Consider for definiteness a β^+-Fermi decay. The angular correlation parameter is $\alpha = +1$ as found from the recoil experiments or as seen in Eq. (6.5) when introducing the values of the coupling constants found in Sect. 21. The physical significance of this statement is that the positron and the neutrino are emitted preferentially in the same direction. Because of the fact that the leptons emitted in an allowed decay are emitted in an s-state (i.e. $L = 0$) the total spin of positron and the neutrino must equal the spin change in the nucleus. This spin change is zero in a Fermi transition. Thus the spins of the two particles must be directed opposite to each other. This in connection with the angular distribution means that when the positron is fully right polarized the neutrino must be left polarized. Proceeding along these lines it can be shown that β^+-decay neutrinos are negatively polarized and β^--decay neutrinos are positively polarized in both Fermi and G-T transitions.

It is seen from this argument that an experimental determination of the neutrino helicity may give us the same information about the coupling constants as that which may be derived from recoil angular correlations. Historically the neutrino helicity investigation of Goldhaber, Grodzins and Sunyar[1] was one of the most decisive clarifications of the nature of the β-decay coupling. The experiment was carried out on the K-capturing 9 hr. isomeric state of Eu^{152}. In principle the experiment is a ν-γ polarization correlation measurement analogous to the β-γ polarization experiments described in Sect. 16. The ν-γ angle is determined by means of resonance scattering. The circular polarization of the γ-ray is measured for the case of 180° between the ν and the γ. Thus what is measured is the induced polarization of the intermediate nuclear state in the direction of emission of the neutrino. Consequently the intermediate nuclear state has to have non-zero spin. The spin sequence of the decay is

$$0^- \xrightarrow{K} 1^- \xrightarrow{\gamma (E1)} 0^+.$$

The cardinal point of the experiment is that of determining the ν-γ-angle by means of resonance scattering. Usually the energy lost by the γ-ray as recoil energy of emitting and absorbing nucleus prevents the resonance scattering. The energy lost by the γ-ray to each of the involved nuclei is

$$\Delta E_\gamma = \frac{E_\gamma^2}{2 M c^2} = 3.2 \text{ ev.} \hspace{3em} (18.1)$$

The kinetic energy of the recoil nucleus left by the neutrino is

$$\Delta E_r = \frac{E_\nu^2}{2 M c^2} = 2.5 \text{ ev.} \hspace{3em} (18.2)$$

---

[1] M. Goldhaber, L. Grodzins and A.W. Sunyar: Phys. Rev. **109**, 1015 (1958).

In these expressions E_γ and E_ν are the energies of the γ-ray and the neutrino respectively and M is the mass of nucleus. In this case the neutrino energy is 840 kv and the γ-energy is 960 kev. The neutrino recoil energy plays an important role in the experiment by causing Doppler shift in the energy of the subsequent γ-ray. A calculation of the shift (stated as increase) in γ-ray energy gives

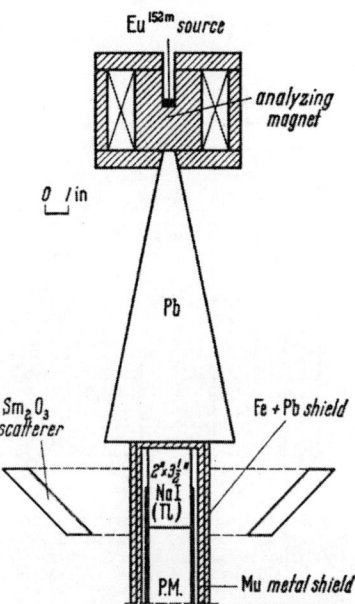

$$\Delta E_\gamma^+ = \Delta E_\nu - \frac{(E_\gamma + E_\nu)^2}{2M c^2} = -8.9 \text{ ev}, \qquad (18.3)$$

$$\Delta E_\gamma^- = \Delta E_\nu - \frac{(E_\gamma - E_\nu)^2}{2M c^2} = +2.5 \text{ ev} \qquad (18.4)$$

where ΔE_γ^+ and ΔE_γ^- are the energy shifts in the cases of the neutrino and the γ-ray emitted in the same direction and in opposite directions respectively. The term ΔE_ν is the neutrino recoil energy which of course has to be disposed of when the γ-decay occurs. The second term is the recoil energy after both transitions $(p_r^2/2M)$. These results are to be compared with the Doppler decrease in the absorption in the absence of the K-capture -3.2 ev. It is seen that the neutrino recoil when the directions are opposite increases the γ-ray energy so that the difference between this and the resonance energy E_0 is only 0.7 ev.

The scattering cross section is described by the Breit-Wigner formula

$$\sigma(E) = \frac{1 + 2J_i}{1 + 2J_f} \cdot \frac{\lambda^2}{8\pi} \frac{\Gamma^2}{\Gamma^2/4 + (E - E_0)^2}. \qquad (18.5)$$

Fig. 81. The arrangement of the neutrino helicity experiment performed by GOLDHABER, GRODZINS and SUNYAR. (After M. GOLDHABER L. GRODZINS and A. W. SUNYAR.)

Consequently the ratio between the cross sections in case of the neutrino and the γ-ray emitted in opposite and same directions is approximately $\frac{12.1^2}{0.7^2} \sim 300$.

Thus the resonance scattering process is indeed very sensitive to the angle between the neutrino and γ-ray momentum directions. The case of Eu^{152} discussed here is especially favourable because of the fact that the neutrino and the γ-ray energies are nearly equal. This fact accommodates the measurements in two ways. The neutrino recoil almost completely compensates for the recoil losses in the emission and absorption processes. This results in the high cross section ratio quoted above and also in a big scattering cross section in the case of opposite neutrino and γ-ray directions.

The apparatus used by GOLDHABER, GRODZINS and SUNYAR (p. 86, Ref. 1) is illustrated in Fig. 81. The circular polarization is measured by the transmission method (see p. 65). The γ-rays leaving the source and magnet part hits a ring-shaped Sm_2O_3 scatterer. Some of the γ-rays are scattered by the Sm_2O_3 into the centrally arranged NaI(Tl) scintillation counter. The pulse height spectrum from the counter is shown in Fig. 82. The lower curve is background radiation originating from Compton-scattering in the scatterer while the two bumps on the upper curve are resonantly scattered γ-radiation. The two bumps are due to the two decay modes of the excited 960 kev level (see insert on Fig. 82). The non-resonant background was measured with the Sm_2O_3 scatterer replaced by a Nd_2O_3 or a lead scatterer containing the same number of atomic electrons replacing the Sm$_2$O$_3$ scatterer. Finally the counting rate in channel B shown in Fig. 82 was measured in the cases of opposite directions of the magnetic field of the transmission magnet. The difference between these counting rates is a measure of the circular polarization of the γ-rays. The relative difference was 0.017 ± 0.003 yielding a (66 ± 15) percent negative polarization of the γ-rays. This is to be compared with the theoretical value of 75% in the actual experiment assuming full negative polarization of the neutrino. Thus the result is well compatible with the assumption of full negative polarization of the neutrinos accompanying K-capture. (It is easily seen that when the directions are opposite the polarization of the neutrino and the γ-ray has to be of the same sign.)

This beautiful experiment is the most direct illustration of the neutrino polarization and is in agreement with the positive and negative polarization of neutrinos accompanying β^--decay and β^+-decay (or K-capture), respectively, inferred from other experiments.

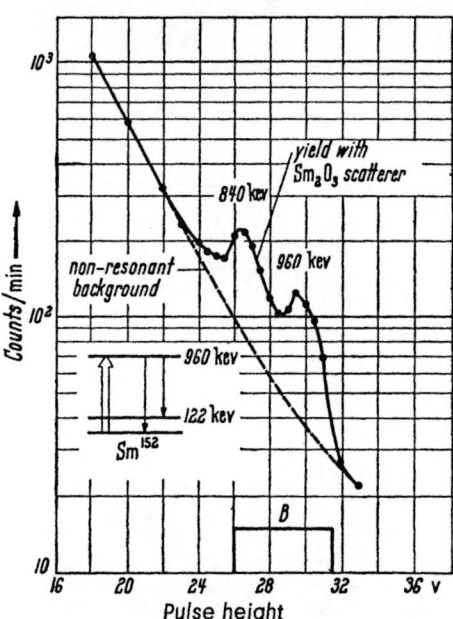

Fig. 82. The pulse height distribution obtained by GOLD-HABER et al. with the instrument shown in Fig. 81. Lower curve is non-resonant background. (After M. GOLDHABER, L. GRODZINS and A. W. SUNYAR.)

E. Coupling constant determination.

19. Introduction. The general theory of β-decay involves a very large number of apriori possible coupling constants. An excellent survey of this topic has been given by PAULI[1]. One set of coupling constants is connected to neutrino emission and one set to antineutrino emission. There are five interactions: Scalar, S, Vector, V, Tensor, T, Axial Vector, A, and Pseudoscalar, P, and each of these interactions may be used in connection with neutrino or antineutrino emission and multiplied by a complex coupling constant i.e. all told 40 coupling constants results. Even though only 35 of these are in principle measurable we shall not attempt to determine all these constants in this very general description of β-decay. Instead we shall adopt the simplifying assumption of a Hamiltonian of the form

$$H = H_S + H_V + H_T + H_A + H_P \tag{19.1}$$

with

$$H_i = \Psi^* O_i \, \Phi \psi^* O_i (C_i + C_i' \gamma_5) \, \varphi + \text{h.c.} \tag{19.2}$$

where Ψ, Φ, ψ and φ are proton, neutron, electron, and neutrino operators respectively and

$$O_S = \gamma_4, \tag{19.3}$$

$$O_V = \gamma_4 \gamma_\mu, \tag{19.4}$$

$$O_A = -i \gamma_4 \gamma_\mu \gamma_5, \tag{19.6}$$

$$O_P = \gamma_4 \gamma_5 \tag{19.7}$$

$$O_T = -\frac{i}{2\sqrt{2}} \gamma_4 (\gamma_\lambda \gamma_\mu - \gamma_\mu \gamma_\lambda), \tag{19.5}$$

and the γ_i's are Dirac matrices.

[1] W. PAULI: Nuovo Cim. 6, 204 (1957); see also T. D. LEE: Proc. Rehovoth Conf. p. 336. Amsterdam: North-Holland Publishing Co. 1958. — A. LUNDBY: Prog. Elementary Particle and Cosmic Ray Physics, 5 p. 4. Amsterdam: North-Holland Publishing Co. 1960.

This Hamiltonian results from the most general theory if lepton conservation is assumed, i.e. if β^--decay is associated with anti-neutrino emission but not with neutrino-emission. Since in this Hamiltonian the coupling constants C_i and C_i' are still complex numbers we are left with essentially 20 unknown constants. It would of course be desirable to deduce from first principles the experimental values of these constants. This can hardly be done at present and instead we shall take refuge to the physisists usual argument of simplicity. We shall simply ask the question: "Can all experimental data be interpreted in terms of one constant", and if the answer is: "No", then we shall ask: "Will two constants suffice" etc.

If the first of these questions is asked then we now know that in order to have parity non-conservation both a C and a C' must be involved. If we continue our series of questions then from the fact that allowed transitions occur for both of the selection rules $\Delta J = 1$ and $\Delta J = 0 \rightarrow 0$ we must conclude that the interaction must involve either scalar or vector components and either tensor or axial vector components. It is customary to call the scalar and vector interactions Fermi (F) interactions and to call the tensor and axial vector interactions Gamow-Teller (G-T) interactions.

Thus we shall now examine the possibility of obtaining a fit to all pertinent experimental data with the assumption of four coupling constants which we may call

$$C_F, C_F', C_{GT} \quad \text{and} \quad C_{GT}'$$

and we shall see that it is possible to obtain a fit with real coupling constants and with

$$C_F = C_F' = C_V \tag{19.8}$$

and

$$C_{GT} = C_{GT}' = C_A \tag{19.9}$$

and with the relative sign of the V and A couplings negative. Because of this result it is purely a convention whether one wants to say that present day β-decay information can be fitted by means of two or four fundamental constants. Furthermore because of the semiempirical nature of the β-decay theory it is an open question whether C_V and C_A are truely constant or whether they may vary to some extent from nucleus to nucleus[1].

On the following pages we shall determine the coupling constants as indicated above and using data obtained by means of the neutron decay and then compare the result thus obtained with supporting evidence from other nuclear transmutations. However, in doing this we shall use the fact that complete polarization of β-particles has been established although this fact has not been demonstrated experimentally on the neutron decay.

20. Polarization of β-particles. The most fundamental type of β-decay experiments is probably the investigation of the polarization of β-decay electrons and positrons. For allowed transitions the polarization expressed in terms of the helicity $\langle \vec{\sigma} \cdot \vec{p} \rangle_{\mp} / |\vec{p}|$ is given by

$$
\begin{aligned}
\xi \langle \vec{\sigma} \cdot \vec{p} \rangle_{\pm} / |\vec{p}| \\
= \left(\pm \frac{2}{1 + \beta/W} \operatorname{Re}\{|\int \vec{\sigma}|^2 (C_T C_T'^* - C_A C_A'^*) + |\int 1|^2 (C_S C_S'^* - C_V C_V'^*)\} + \right. \\
\left. + \frac{2}{1 + \beta/W} \operatorname{Im}\{|\int \vec{\sigma}|^2 (C_A C_T'^* + C_A' C_T^*) + |\int 1|^2 (C_V C_S'^* + C_V' C_S^*)\} \frac{\alpha Z}{p} \right) \frac{v}{c}
\end{aligned} \tag{20.1}
$$

[1] A. WINTHER: Thesis, University of Copenhagen 1961.

with

$$\xi = |\textstyle\int \vec{\sigma}|^2(|C_T|^2 + |C_T'|^2 + |C_A|^2 + |C_A'|^2) + \atop + |\textstyle\int 1|^2(|C_S|^2 + |C_S'|^2 + |C_V|^2 + |C_V'|^2)} \qquad (20.2)$$

and

$$\beta\xi = \gamma|\textstyle\int \vec{\sigma}|^2(C_T^* C_A + C_T'^* C_A' + C_T C_A^* + C_T' C_A'^*) + \atop + \gamma|\textstyle\int 1|^2(C_S^* C_V + C_S'^* C_V' + C_S C_V^* + C_S' C_V'^*)} \qquad (20.3)$$

with

$$\gamma = [1 - (\alpha Z)^2]^{\frac{1}{2}}. \qquad (20.4)$$

It is seen from this expression that the experimental result of full polarization for allowed transitions, for all β-particle energies, and with electrons left-handed and positrons right-handed leads to the conclusion in our approach that

$$\left. \begin{aligned} C_T &= -C_T', \\ C_S &= -C_S' \\ C_A &= +C_A', \\ C_V &= +C_V'. \end{aligned} \right\} \qquad (20.5)$$

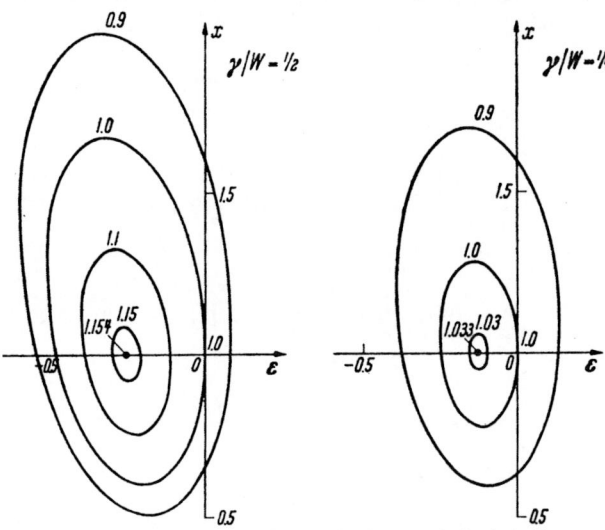

Fig. 83. The figure gives an altitude chart for the numerical polarization as a function of x and ε.

From this it follows that the Fierz interference term vanishes. Strictly speaking this conclusion can be drawn only in the exact limit of full v/c polarization. It should be mentioned that the same information about the G-T coupling is obtained from the main term in Eq. (15.2) and Eq. (16.4), that is from the β-distribution from polarized nuclei and the β-γ-polarization correlation. It is quite illustrative for the understanding of the immense complication involved in a strictly logical deduction of the values of the 20 coupling constants to consider this question in slightly greater detail. Thus let us return to the more complicated case of small admixtures of tensor interaction in a pure G-T transition dominated by the axial vector coupling. Furthermore for simplicity let us ignore the $\alpha Z/p$ term, and let us assume that we are dealing with β-particles of such an energy that $\gamma/W = \frac{1}{2}$. Furthermore for simplicity let us adopt the picture of real coupling constants and $C_T = C_T' = \varepsilon C_A$ and let us put $C_A' = x C_A$. We then find the helicity for electron emission

$$\langle \vec{\sigma} \cdot \vec{p} \rangle/|\vec{p}| = \frac{2\varepsilon^2 - 2x}{2\varepsilon^2 + 1 + x^2 + \varepsilon + \varepsilon x}. \qquad (20.6)$$

This example corresponds to going one step further in our series of questions dictated by the principle of simplicity. The permissible limit on ε yields an idea of the uncertainty in our procedure as regards the results that would appear if a strictly logical evaluation of the coupling constants was attempted. However, the complete scheme involving 20 constants would of course be much more complicated. Eq. (20.6) can be considered a function of ε and x and in Fig. 83 this function is illustrated together with the equivalent expression for $\gamma/W = \frac{1}{4}$.

It is evident from these diagrams that an overall uncertainty of the helicity of a few percent mean relatively large uncertainties in the coupling constants if we want to investigate the effects from the possibility that the "small" coupling constants (scalar and tensor) are not negligible. For the following derivation of numerical values for the coupling constants we shall however neglect such effects entirely.

21. The neutron decay. From the measurement of the constant D for the neutron decay we may conclude under our assumption of only one Fermi and one Gamow-Teller interaction that the coupling constants are real numbers. This may be seen from the expression for D which considering Eqs. (20.5) may be written

$$D = \frac{2 \operatorname{Im} (C_S C_T^* - C_V C_A^*)}{|C_S|^2 + |C_V|^2 + 3(|C_A|^2 + |C_T|^2)} = 0.04 \pm 0.05 \tag{21.1}$$

since for the neutron decay $|\int 1|^2 = 1$ and $|\int \vec{\sigma}|^2 = 3$ and since we may neglect Fierz interference. From this result we may conclude that the phase of C_F and C_{GT} is smaller than $\pm 8° + n\pi$ and consequently we shall neglect the possible imaginary part of the coupling constants entirely. Also $C_S = C_A = 0$ or $C_T = C_V = 0$ would mean $D = 0$. These possibilities are ruled out because they require numerical equality of A and B [see Eqs. (21.2) to (21.3)].

With real coupling constants and with the result of Eqs. (20.5) we now have for the constants A and B for the neutron decay

$$A = - \frac{2(C_T^2 + C_A^2 + C_T C_S + C_A C_V)}{C_S^2 + C_V^2 + 3(C_A^2 + C_T^2)} = - 0.114 \pm 0.019, \tag{21.2}$$

$$B = \frac{2(C_A^2 - C_T^2 - C_V C_A + C_S C_T)}{C_S^2 + C_V^2 + 3(C_A^2 + C_T^2)} = 0.88 \pm 0.15. \tag{21.3}$$

These two results are compatible with the A-V coupling only and yield

$$C_A = - (1.25 \pm 0.05) C_V. \tag{21.4}$$

We can now use the ft value for the neutron decay for obtaining the absolute magnitude of the decay constants and in this way one finds

$$B = ft \left[(1 - x) | \int 1 |^2 + x | \int \vec{\sigma} |^2 \right] = 2740 \pm 140 \tag{21.5}$$

with

$$x = \frac{C_T^2 + C_A^2}{C_S^2 + C_V^2 + C_T^2 + C_A^2} = \frac{C_A^2}{C_V^2 + C_A^2} = 0.61 \pm 0.02 \tag{21.6}$$

and from [see Eq. (2.16)]

$$B = \frac{\pi^3 \hbar^7 \ln 2}{(C_V^2 + C_A^2) m^5 c^4} \tag{21.7}$$

we get

$$\sqrt{2} |C_V| \sim |g_V| = (1.32 \pm 0.07) \times 10^{-49} \text{ erg cm}^3. \tag{21.8}$$

The narrow limits of error of Eq. (21.8) are of course not valid for $|C_V|$ because of the inaccuracy of Eqs. (20.5).

Finally we should mention also the result for the angular correlation coefficient α which results in the equation

$$\frac{C_V^2 - C_A^2}{C_V^2 + 3 C_A^2} = - 0.09 \pm 0.11 \tag{21.9}$$

which yields

$$x = 0.42 \pm 0.10 \tag{21.10}$$

in reasonable agreement with the value derived from A and B and given above.

It is seen from the discussion so far that in principle the β-decay interaction can be determined from the neutron decay alone. However, much additional evidence can be derived from other experiments and historically the neutron decay has played a less decisive role. For this reason we shall briefly mention the most important conclusions which have been drawn from nuclear β-decay.

22. Supporting evidence. The final decision of the β-decay coupling being V-A was very much facilitated by the neutrino helicity experiment carried out by Goldhaber, Grodzins and Sunyar[1]. In this experiment it was demonstrated that the neutrino helicity is negative in the $0^- \rightarrow 1^-$ K-capture of Eu^{152m}. This

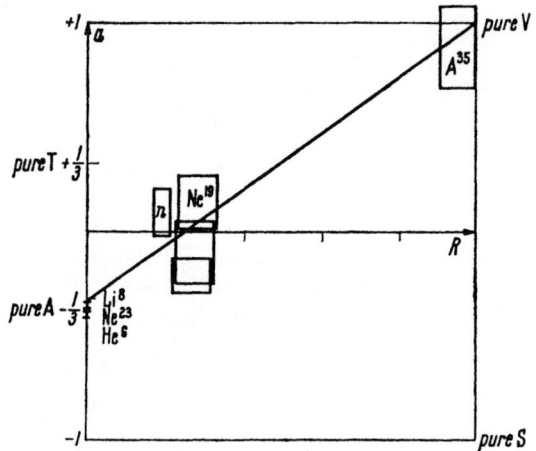

Fig. 84. The β-ν angular correlation parameter α plotted as a function of

$$R = \frac{(|C_S|^2 + |C'_S|^2 + |C_V|^2 + |C'_V|^2)\,|\int 1|^2}{(|C_S|^2 + |C'_S|^2 + |C_V|^2 + |C'_V|^2)\,|\int 1|^2 + (|C_T|^2 + |C'_T|^2 + |C_A|^2 + |C'_A|^2)\,|\int \vec{\sigma}|^2}$$

in a so-called Scott diagram. (After A. Lundby.)

result is incompatible with tensor interaction and agrees with axial vector interaction.

The result from recoil angular correlation experiments are usually enterpreted in the so called Scott diagram. In this diagram one plots α as a function of R with

$$\alpha = \frac{(C_V^2 - C_S^2)\,|\int 1|^2 + \frac{1}{3}(C_T^2 - C_A^2)\,|\int \vec{\sigma}|^2}{(C_S^2 + C_V^2)\,|\int 1|^2 + (C_T^2 + C_A^2)\,|\int \vec{\sigma}|^2} = R \cdot \alpha_F + (1 - R)\,\alpha_{GT} \tag{22.1}$$

where

$$R = \frac{|\int 1|^2}{|\int 1|^2 + \dfrac{x}{1-x}\,|\int \vec{\sigma}|^2} \tag{22.2}$$

and α_F is the α for pure Fermi transition and α_{GT} that corresponding to pure G-T-transitions. This plot is drawn in order to distinguish between the VA and ST combinations. The present results given in Sect. 6 are collected in the Scott diagram in Fig. 84 and is clearly seen to support the V-A coupling.

For the relative values of the F and GT coupling constants the mirror nuclei ft values have been of great importance.

[1] M. Goldhaber, L. Grodzins and A. W. Sunyar: Phys. Rev. **109**, 1015 (1958). This experiment is discussed in Sect. 18.

If the matrix elements for a decay are known then a plot of B vs. x yields a straight line for each such decay[1] where

$$B = ft \left[(1 - x) \mid \int 1 \mid^2 + x \mid \int \vec{\sigma} \mid^2 \right] = \frac{\pi^3 \hbar^7 \ln 2}{(C_V^2 + C_A^2) \, m^5 \, c^4} \qquad (22.3)$$

and with

$$x = \frac{C_A^2}{C_A^2 + C_V^2}. \qquad (22.4)$$

The common intersection point of such lines provide a determination of B and x, or if B and x and $\mid \int 1 \mid^2$ are known the ft value permits a determination of $\mid \int \vec{\sigma} \mid^2$, a procedure which has been utilized in connection with the construction of the Scott diagram.

Usually the Fermi matrix elements are determined from

$$\mid \int 1 \mid^2 = [T_1(T_1 + 1) - T_{1Z} T_{2Z}] \, \delta_{T_1 T_2} \qquad (22.5)$$

where T_1 and T_2 are the isotopic spins of the nuclear states in question. T_{1Z} and T_{2Z} are their z-components, and δ_{ik} is the Kronecker symbol. The formula is derived from the assumption of charge independence of nuclear forces only[2]. The Coulomb field between the protons will introduce a correction to this expression. This correction has been calculated by McDonald[3] and is found to be negligible for the light nuclei. Thus the selection rule implied by the $\delta_{T_1 T_2}$ is not strict in the region of heavier nuclides.

The matrix elements for the free neutron are the most reliable matrix

Table 7. *Data of interest for the B-x diagram shown in Fig. 85.*

Decay	ft	Reference	$\mid \int 1 \mid^2$	$\mid \int \vec{\sigma} \mid^2$
n	1185 ± 40	4	1	3
O^{14}	3061 ± 10	5	2	
Al^{26}	3070 ± 60	6, 7	2	
Cl^{34}	3110 ± 70	6	2	
H^3	1132 ± 40	8	1	3.62 ± 0.10
O^{15}	4380 ± 100	7	1	0.350
F^{17}	2360 ± 80	7	1	1.373
Ca^{39}	4300 ± 100	9	1	0.390

elements obtainable and they have been used extensively in the section on the neutron decay. The second best matrix elements are obtained for the $0 \to 0$ transitions among pure isotopic spin states. The important decays are those of O^{14}, Al^{26} and Cl^{34} for which recent data are given in Table 7. For these decays one finds $\mid \int 1 \mid^2 = 2$ and $\mid \int \vec{\sigma} \mid^2 = 0$. In Fig. 85 the B-x line for the neutron decay and the average line for the $0 \to 0$ transitions are shown.

From the intersection point one finds $B = 2570 \pm 25$ and $x = 0.583 \pm 0.005$ in agreement with the determination from the experiments with polarized neutrons. The $0 \to 0$ transitions permit a determination of the absolute magnitude of C_V. The result is

$$\sqrt{2} \mid C_V \mid = \mid g_V \mid = (1.409 \pm 0.007) \times 10^{-4} \text{ erg cm}^3 \qquad (22.6)$$

in reasonable agreement with the neutron result.

[1] See e.g. O. Kofoed-Hansen and A. Winther: Phys. Rev. **86**, 428 (1952). — Matt. Fys. Medd. Dan. Vid. Selskab. **27**, No. 14 (1953); **30**, No. 20 (1956).

[2] E. Wigner and E. Feenberg: Rep. Progr. Phys. **8**, 274 (1941).

[3] W.M. McDonald: Thesis, Princeton 1955.

[4] See Sect. 17.

[5] D.L. Hendrie and J.B. Gerhart: Phys. Rev. **121**, 846 (1961).

[6] C. van der Leun: Thesis, Utrecht 1958.

[7] For references, see O. Kofoed-Hansen and A. Winther: Dan. mat.-fys. Medd. **30**, No. 20 (1956).

[8] See M. Goldhaber: 1958 Annual Internat. Conf. on High Energy Phys. at CERN, Rep. p. 233 (1958).

[9] O.C. Kistner and B.M. Rustad: Phys. Rev. **112**, 1972 (1958). — J.E. Cline and P.R. Chagnon: Bull. Amer. Phys. Soc., Ser. II **3**, 206 (1958).

For the mirror transitions of nuclei of closed shells $\pm$ one nucleon it is to be believed that the single particle model gives reasonably good values for $|\int \vec{\sigma}|^2$ and that furthermore the deviation of the measured magnetic moments from the Schmidt values permits the calculation of a correction to these matrix elements so that reasonable results may be obtained. These semiempirical matrix elements are given by the expression

$$|\textstyle\int \vec{\sigma}|^2 = 4\,\frac{I+1}{I}\,\frac{(\mu - g_l I)^2}{(g_s - g_l)^2} \tag{22.7}$$

where g_l and g_s are the gyromagnetic ratios for orbital angular momentum and spin of the odd particle. At the same time $|\int 1|^2 = 1$. The data which form a

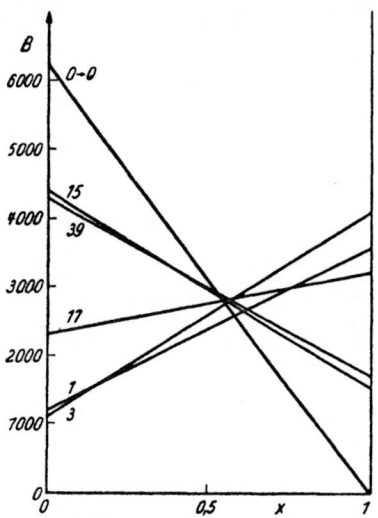

Fig. 85. B-x diagram for the neutron decay, $0 \to 0$ transitions and selected mirror transitions.

basis for the B-x lines of these decays are given in Table 7 also and the lines are shown in Fig. 85. The internal consistency of the common B-x point for these decays and for the $0 \to 0$ transitions is deceivingly good, but does not agree too well with the value for the neutron decay. The results are $B = 2820 \pm 50$ and $x = 0.55 \pm 0.01$. The formula Eq. (22.7) is derived in a primitive manner and can be seriously critizised[1], one of the major objections being that it predicts $|\int \vec{\sigma}|^2 > 3$ for the tritium decay.

In spite of the small disagreements pointed out above the general agreement with the result for the neutron decay is quite striking for all of the supporting evidence mentioned in this section. Of course it should be remembered that complex nuclei are not elementary structures and many corrections are to be expected. Thus Coulomb corrections, mean deviations from isotopic spin as an exact quantum number[2], and relativistic and mesonic corrections are to be expected[3].

Before leaving this section some conclusions discussed in Sects. 11 and 12 should be mentioned again.

It is not found to be possible to say anything about the Fierz interference due to the recently found deviations from the statistical shape which can not be interpreted as Fierz interference.

The time reversal invariance was found to be proved also from the RaE-experiments.

The relative sign of g_A and g_V was found to be negative from the β-γ angular correlation in Ba[139].

Finally, it might also be of interest to remember that the shape of allowed β-spectra like that of He[6] may be used for setting a limit on the pseudoscalar coupling constant. Unfortunately, this limit is so high $\sim 50 C_A$ that it is of little interest in the basic understanding of the Fermi interaction.

---

[1] See e.g. J.M. BLATT: Phys. Rev. **89**, 86 (1953). — M. BOLSTERLI and E. FEENBERG: Phys. Rev. **97**, 736 (1955). — J.S. BELL and R.J. BLIN-STOYLE: AERE T/P 35.

[2] W.M. McDONALD: Bull. Amer. Phys. Soc., Ser. II **3**, 48 (1958). — Phys. Rev. **110**, 1420 (1958).

[3] A. WINTHER: Thesis, University of Copenhagen 1961 (in press).

23. Concluding remarks. In the above most of the fundamental evidence on the β-decay interaction has been presented. The interpretation has been guided by the assumption that two coupling constants only are necessary and that all other coupling constants are small enough to be negligible for this analysis. The result is that the coupling is A-V and that $C_A \simeq -1.25 C_V$. Much speculation is devoted to this 25% deviation from $C_A = C_V$. On the other hand the entire picture is not rigorously consistent as may be seen from the fact that the division of the coupling constants into dominating and negligible constants is only good to statements like this: "The negligible constants are smaller than 25% of the dominating constants" as shown by the analysis of the electron polarization results (see Sect. 20).

F. Electron capture.

24. *K* capture. The primary transition in electron capture consists in the absorption of an orbital electron by the nucleus and the simultaneous emission of a neutrino. The only detectable phenomenon from this primary transition is the motion of the recoil. The recoil experiments on K capture have been discussed in Sect. 6. These experiments can hardly be said to serve as methods for the detection of K capture. All detection methods for K capture must therefore rely on secondary effects which are associated with the *charge change* of the nucleus, the K *(L, M* etc.*) vacancy* created in the recoil atom, or higher order effects such as *internal bremsstrahlung*.

The charge change in the nucleus can be detected by chemical methods especially if the daughter nucleus is radioactive and tracer chemistry can be applied. The charge change also manifests itself in the X-rays associated with the filling of the $K(L, M$ etc.) shell vacancies. Finally, the decay may lead to excited states or radioactive daughter with a characteristic decay scheme that will identify the daughter. If radioactive decay or sufficiently long-lived isomeric decay occurs in the daughter, the rate of growth of the daughter activity can be used as a means of investigating the K capture.

The $K(L, M$ etc.) shell vacancies may be filled by X-ray emission or by *Auger effect*. As mentioned, the X-rays are those characteristic for the daughter atom. ALVAREZ[1] utilized this fact in demonstrating K capture in the decays of V^{48} and Ga^{67}. Both of these decays emit positrons also, but practically no conversion electrons. If a nuclear reaction leads to excited nuclear states which decay by internal conversion, characteristic X-rays are, of course, found from the daughter. This must be taken into account in order not to confuse an isomeric transition with K capture. K capture itself often leads to such excited states in the daughter and a complete decay scheme is desirable if one wants to evaluate the number of K shell vacancies actually created in the K capture event. In evaluating this number the Auger effect should also be taken into account. The necessary information is furnished by the fluorescent yield for the atom in question. Fluorescent yield curves have been given in the literature[2].

In older experiments the characteristic X-rays were identified by absorption measurements, possibly using the selective absorption of a material with a suitably discriminative K shell absorption edge. Now such techniques are usually replaced by proportional counter spectrometers. The Auger electrons may be studied in electron spectrometers. Electron capture has now been found in all positron emitters of large enough Z to give reasonable ease of detection and also for many

---

[1] L. W. ALVAREZ: Phys. Rev. **54**, 486 (1938).
[2] C. D. BROYLES, D. A. THOMAS and S. K. HAYNES: Phys. Rev. **89**, 715 (1953).

cases where positron emission is energetically impossible. A review of K capture investigations has been given by Major and Biedenharn[1].

It is usually quite difficult to carry out precise determinations of the number of K capture events occurring in a source because of the difficulties in interpreting the number of X-rays observed and also because of the uncertainty in the fluorescent yield. The ratio of captures to positrons is, however, very useful for the study of β-decay, especially as regards the Fierz interference terms. With this problem in mind the decay of Na^{22} has been studied by Sherr and Miller[2]. The decay scheme is illustrated in Fig. 86. In Na^{22} the decay leads to an excited state in Ne^{22}, and the total number of decays can be found from the number of 1.28 Mev γ-rays. This can then be compared with the total number of positrons in order to give the number of capture events. Electron capture was found to occur in (9.9 ± 0.6)% of the transitions and from this it follows that the ratio R of captures to positron emissions is $R = 0.110 \pm 0.006$. This ratio can be used to give a value for the Fierz interference term in the G-T interaction since the selection rules imply that only the G-T interaction is present in the allowed Na^{22} decay. Let R_0 be the theoretical ratio in the absence of any Fierz interference. It then follows that β_{GT} can be found from the expression

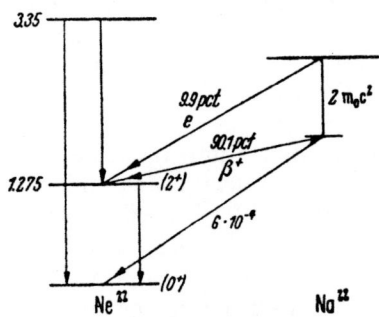

Fig. 86. The decay scheme of Na^{22}.

$$\frac{R}{R_0} = \frac{1 + \beta_{GT}}{1 - \left\langle \dfrac{1}{W} \right\rangle \beta_{GT}} \qquad (24.1)$$

where $\langle 1/W \rangle$ is the mean value of $1/W$ for the positron spectrum. The above result for R leads to $\beta_{GT} = -0.02 \pm 0.04$. This result has already been mentioned in Sect. 12. It represents the lowest limit which can be put to the Fierz interference. It is very much to be hoped that similar experiments may be carried out on a pure Fermi transition. As shown in Sect. C, however, the Fermi interaction may occur mainly among very light elements where the capture-to-positron ratio is expected to be very small. The experiment will consequently be more difficult.

25. L/K capture ratios. The theory of electron capture has been given by Marshak, by Rose and Jackson and by Brysk and Rose[3]. These authors have calculated K and L capture probabilities.

The measurement of L/K *capture ratios* combined with the knowledge of the energy release in the decay is a good means of testing the theory because nuclear matrix elements drop out. A review of experimental material has been given by Robinson and Fink[4]. Here we shall briefly sketch the most important methods used for obtaining L/K capture ratios. Experimentally, we would define a K capture reaction as a decay in which immediately after the decay a K shell vacancy is left in the recoil, and an L capture is a reaction where an L shell vacancy is left in the recoil. The total energy released by the recoil atom before this atom reaches its ground state will therefore correspond to the energy value of the K shell absorption edge and the L shell absorption edge in the two

[1] J. K. Major and L. C. Biedenharn: Rev. Mod. Phys. **26**, 321 (1954).
[2] R. Sherr and R. H. Miller: Phys. Rev. **93**, 1076 (1954).
[3] R. E. Marshak: Phys. Rev. **61**, 431 (1942). — M. E. Rose and J. L. Jackson: Phys. Rev. **76**, 1540 (1949). — H. Brysk and M. E. Rose: Oak Ridge National Laboratory Report, No. 1830, 1955, and Rev. Mod. Phys. **30**, 1169 (1958).
[4] B. L. Robinson and R. W. Fink: Rev. Mod. Phys. **27**, 424 (1955); **32**, 117 (1960).

cases, respectively. If, therefore, the radioactivity is immersed in a substance in such a manner that all energy is absorbed and registered, the two types of events will be distinguishable by the amount of energy released by the atom. The times involved in the atomic de-excitations in question are short enough to permit the detection of all the energy simultaneously rather than as energy of individual quanta or Auger electrons. The most reliable results are consequently obtained when the radioactivity is part of the filling gas of a proportional counter or is grown into a scintillation crystal.

One of the most carefully studied cases is the decay of A^{37}. *L* capture was demonstrated for this isotope by Pontecorvo, Kirwood and Hanna[1]. Very careful measurements have also been performed by Langevin and Radvanyi[2]. In both of these investigations A^{37} was introduced into a proportional counter of such dimensions that X-rays could be practically completely absorbed inside the counter and peaks corresponding to *K* and *L* shell vacancies could be detected. The areas of the peaks give $L/K = 0.092 \pm 0.010$, in very good agreement with theoretical predictions of $L/K = 0.082$. Similar methods have been applied to other isotopes also.

Scintillation counters do not in general permit a detection of the *L* shell vacancy de-excitation because of the relatively high energy expenditure per photon produced in the detector. However, the *K* shell vacancies can be detected, and in cases where the decay leads to an excited state in the daughter the total number of decays can be found from the number of γ-rays from de-excitation of the daughter. The number of *L* and higher shell conversions can then be found by subtraction of the two results. In cases where the de-excitation of the daughter is prompt two peaks will occur, one corresponding to the γ-ray energy + the *K* energy, and one corresponding to the γ-ray energy + the *L* energy, and the separation can again be made in favourable cases. The scintillation crystal must, of course, be large enough to prohibit the escape of secondary quanta. Experiments of this kind have been performed by der Mateosian[3] who has grown crystals containing Cd^{109} and I^{125}. In the case of Cd^{109} the daughter state is metastable with a half-life of 39 sec and the first method is used, in the case of I^{125} the daughter decays by promt decay and the second method is applied.

Both of the methods mentioned so far have the distinct advantage that the *K* shell vacancies will give rise to a full *K* energy peak and correspondingly for the *L* shell vacancies. When the radioactive source is kept outside the detector, the $K - L$ X-ray may miss the detector and the $L - M$ X-ray be detected, thereby masking the $L - M$ X-rays from primary *L* capture. Experiments with external sources have also been performed, but the interpretation of the data is complicated because of effects of the type just mentioned and because of the necessity for geometry and efficiency corrections for the instrument. Window absorption and possible scattering must also be taken into account. No very reliable results have been obtained by such methods.

The total number of reliable *L/K* capture ratios measured is relatively small. The above mentioned investigations of A^{37} constitute the most careful work and give the closest agreement with the theory.

Before we leave the subject of *L*-capture it should be mentioned that in the case of Be^{7} the *L* electrons constitute the valence electrons. Thus the half-life of Be^{7} can be influenced by the chemical binding. Segrè and coworkers, and

[1] B. Pontecorvo, D. H. W. Kirwood and G. C. Hanna: Phys. Rev. **75**, 982 (1949).
[2] M. Langevin and P. Radvanyi: C. R. Acad. Sci., Paris **241**, 33 (1955).
[3] E. der Mateosian: Phys. Rev. **92**, 938 (1953).

Kraushaar *et al.*[1] have used pairs of balanced ion chambers of the type shown in Fig. 4 in order to detect the small change in half-life when Be is used in different chemical forms. They find that Be^7 in BeF_2 decays about 0.07% faster than in Be metal. This experiment, of course, adds to our confidence in our description of electron capture.

26. Energy determination in electron capture. The energy release in K capture can be determined in several indirect ways. The energy release is connected with the atomic masses in question by the equation [cf. Eqs. (2.9) and (2.10)]

$$W_0^K = (M_{A,Z} - M_{A,Z-1}) - B_K = \Delta - B_K, \qquad (26.1)$$

where W_0^K is the neutrino energy apart from the negligible recoil energy and B_K is the binding energy of the K electron.

In the case of positron emission, W_0^K can be found from Eqs. (2.10) and (26.1) provided the positron maximum energy has been measured.

Whenever nuclear reaction data, especially (p, n) thresholds are available, the K capture energy may be found from Eq. (26.1). An example is provided by A^{37} mentioned in the discussion of K capture recoil experiments (Sect. 6).

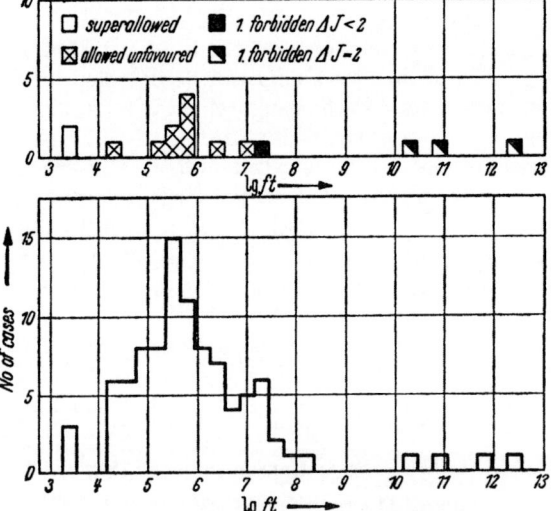

Fig. 87. $\log_{10} ft$ histogram for electron capture.

L/K capture ratios may be used for the energy determination, especially when $W_0^K \approx B_K$. This ratio is proportional to the square of the ratio $(\Delta - B_L)/(\Delta - B_K)$ which varies rapidly when W_0^K is of the same order of magnitude or smaller than B_K. Examples of energies found in this way are provided by the decays of Cd^{109} and I^{125} mentioned in Sect. 25.

Internal bremsstrahlung spectra may also be used for the energy determinations. Although the lower part of the spectrum is somewhat in doubt (cf. Sect. 28β), there is no doubt that near the maximum energy the bremsstrahlung spectrum varies as

$$P(k) = k(W_0^K - k)^2, \qquad (26.2)$$

where k is the photon energy. Consequently, from a measured bremsstrahlung spectrum W_0^K may be found. Examples of energy determinations by this procedure are found, e.g., in the decays of A^{37}, V^{49} and Fe^{55}.

Finally, in a few cases the energy release has been measured by recoil investigations (cf. Sect. 6).

Energy and half-lives for K capture transitions lead to ft values in which the "Fermi integral" is now given by

$$f = \frac{\pi}{2}(W_0^K)^2 \frac{(1+\gamma)R^{2\gamma-2}(2\alpha Z)^{2\gamma+1}}{2\Gamma(2\gamma+1)} \qquad (26.3)$$

[1] E. Segrè and C. E. Wiegand: Phys. Rev. 75, 39 (1949); 81, 284 (1951). — R. F. Leininger, E. Segrè and C. E. Wiegand: Phys. Rev. 76, 897 (1949); 81, 280 (1951). — J. J. Kraushaar, E. D. Wilson and K. T. Bainbridge: Phys. Rev. 90, 610 (1953).

where the symbols have the same meaning as in Eqs. (2.6) to (2.8). Numerical values for the estimate of K capture ft values are included in Figs. 6 and 9.

In Fig. 87 we show histograms of ft values for K capture. In the upper diagram we show those of the cases listed by KING [5] for which the spin assignments are reasonably certain. In the lower curve we show the histogram of ft values given in the tables of MAJOR and BIEDENHARN[1]. It is evident from Fig. 87 that no striking differences exist between the $\log_{10} ft$ distribution for K capture and for β-decay. Relatively few highly forbidden K captures are known. This is undoubtedly due to the difficulties in measuring and identifying electron capture events.

Summing up, we may say that the present situation in K capture investigations is in a relatively satisfactory state when compared with the theory of β-decay, but that more work is desirable both as regards the measurements of capture to positron ratios and as regards the determinations of the energy release and the measurements of K/L capture ratios.

G. Related reactions.

27. Double β-decay. Double β-decay is a pure second order β-decay reaction. It is expected to occur between non-neighbouring stable isobars of two units difference in Z. In Fig. 88 we show two isobaric sections in the nuclear energy valley for even nuclei. The diagrams are meant only as illustrations and do not

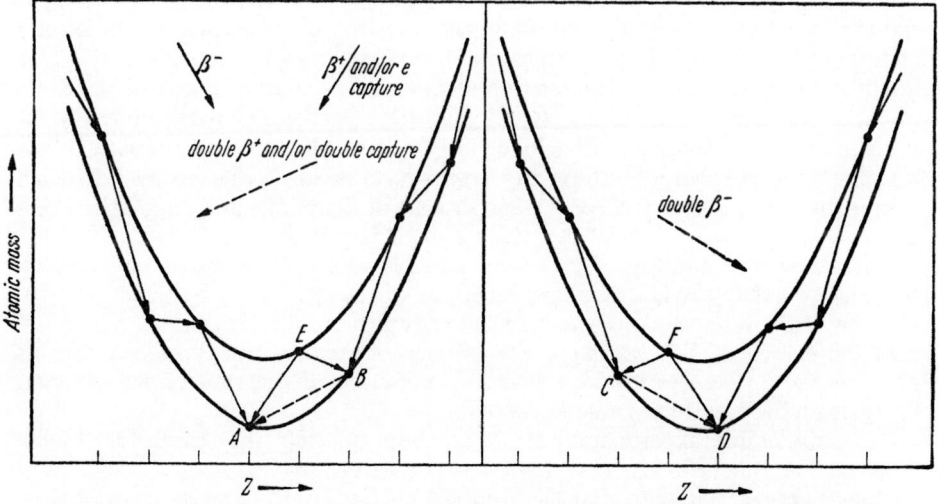

Fig. 88. Two sections in the energy valley of nuclei for even isobars illustrating the double β-decay reaction.

refer to any actual cases. It is evident from these diagrams that nucleus B may be expected to be unstable against decay into nucleus A by double β^+-decay and/or double electron capture, and that nucleus C may decay into nucleus D by double β^--decay. Double β-decay may occur according to two different schemes, depending on the nature of the neutrino. The neutrino emitted (virtually) in the first step of the reaction may be absorbed in the second step, i.e.,

$$C \to F + \beta^- + \nu \to D + \beta^- + \beta^-,   (27.1)$$

where the intermediate state is a virtual state used in the second-order calculation.

---

[1] J. K. MAJOR and L. C. BIEDENHARN: Rev. Mod. Phys. **26**, 321 (1954).

If the (virtual) neutrino emitted in the first step is not reabsorbed another neutrino has to be emitted in the second stop, i.e.

$$C \to F + \beta^- + \nu \to D + \beta^- + \beta^- + \nu + \nu, \tag{27.2}$$

where again the intermediate step is a virtual state. In the case of reaction (27.1) the sum of the electron kinetic energies equals the energy release in the reaction. In the case (27.2) this sum may have all values from zero to the energy release in the reaction. Similar considerations apply to double β^+-decay with the exception that here energy corresponding to four electron rest masses must be provided [cf. Eq. (2.10)]. For double K capture, neutrinos have to be emitted in order to provide a means of carrying away the energy released.

The theoretical significance of distinguishing between these two processes has been changed with the breakdown of parity conservation. As long as parity is conserved the difference between the processes is the following: A necessary condition for a neutrino participating in reaction (27.1) is that the one emitted in the first step (β^--decay process) can reverse the β^--decay process. This can be expressed with other words: that neutrinos and antineutrinos are identical, that is they are Majorana particles and not Dirac particles.

Meanwhile the problem has to be reconsidered with present day β-decay theory in mind. We can again use the argument ruling out the process (7.7) in Sect. 7 concerning direct neutrino detection. Today it is a well established fact that the β-particles and neutrinos emitted during a β-decay are fully polarized. Applied to the reaction (27.1) this means that the neutrino emitted in the first step is righthanded (forwards polarized) while the neutrino participating in the second step must be lefthanded. This rules out the possibility of reaction (27.1). This argument is quite independent on the somewhat speculative adoption of one of the pictures of the neutrino: (1) as Majorana particles showing chirality invariance, or (2) as Dirac particles following the law of lepton conservation. What remains is the fact that it is interesting to seek reaction (27.2) the existence of which is predicted from β-decay theory. The absence of (27.1) means simply that there are at least two not identical neutrinos (which fact is already known).

If reactions of the type (27.1) were possible the half-life would be expected to be shorter than the half-lives for reactions of the type (27.2) by a factor $\sim 10^8$ because of the large contribution to the decay rate from the sum over all virtual neutrino energies. The two major distinctions between the two types of reactions are thus the differences in the spectral distribution of the sum of the electron energies and the half-life to be expected.

A considerable experimental effort has been put into the search for double β-decay. The theory of reaction (27.2) has been given by Goeppert-Mayer[1] and the theory of reaction (27.1) has been worked out by Furry[1] and by Primakoff[1]. The theory of double capture and of single positron emission has been given by Winter[1]. These theories are based on the assumption of parity conservation. Because of the smallness of the coupling constants in Eq. (21.8) the half-lives to be expected for these reactions are extremely long, namely, $\sim 10^{21}$ years for reactions (27.2) and $\sim 10^{13}$ years for reaction (27.1) in cases of superallowed transition with several Mev energy release. However, none of the cases occurring in nature are expected to yield superallowed transitions, and a product of two unfavoured factors is expected to increase the lifetime by a factor of $\sim 10^2$ to 10^{11} where these figures have been obtained by a comparison of superallowed ft values with the extremes of

---

[1] W. H. Furry: Phys. Rev. **56**, 1184 (1939). — H. Primakoff: Phys. Rev. **85**, 888 (1952). M. Goeppert-Mayer: Phys. Rev. **48**, 512 (1935). — R. G. Winter: Phys. Rev. **100**, 142 (1955).

the values for unfavoured transitions in Fig. 39. Consequently, even half-lives of $\sim 10^{20}$ years may be compatible with the reaction scheme (27.1) under unfavourable conditions. In such cases where both the electron and positron decay ft values of nucleus F are known, more can be said about the possible limit on the double β-decay half-life.

The long half-life renders the experimental search for double β-decay extremely difficult. Even with gram quantities of the isotope in question and essentially 4π counting it is possible to hope for only a few decays per day. Consequently, coincidence technique has to be used to diminish background effects. The experimental arrangement consists essentially of two counters between which the source is placed. Even then difficulties arise. Each of the two electrons comprising the double β-decay and thus emitted from the same source point must necessarily be permitted to reach its own counter. No more shielding can be placed between the counters at the source position than that which the electrons can penetrate. Thus the shielding between the counters can never be very effective and the rate of accidental coincidences becomes high. Anticoincidence shields against cosmic ray background have to be used and the source in question must be compared with a dummy source of an isotope that is not expected to show double β-decay.

The experimental data are somewhat conflicting. Several investigations have set lower limits of 10^6 to 10^{18} years for the half-lives of expected double β-decaying isotopes[1]. Only two investigations indicate positive results. INGHRAM and REYNOLDS[2] have reported an excess of Xe^{130} in the mass spectrum of xenon extracted from tellurium containing ore. This could be accounted for by the assumption of a 10^{21} years half-life of Te^{130} with respect to double β-decay. The result is uncertain due to the difficulties involved in a quantitative extraction of xenon from the ore and due to questionable long time retention of xenon by the rock. The lifetime seems somewhat long for a reaction of the type (27.1). McCARTHY[3] has reported the finding of coincident electrons from Ca^{48} with the expected energy of ~ 4 Mev. This would support the decay scheme (27.1). The lifetime given by McCARTHY $\sim 1.6 \times 10^{17}$ years would also be reasonable for this type of decay. In Zr^{96} McCARTHY found a possible double β-decay effect corresponding to a half-life of $\sim 6 \times 10^{16}$ years. Very recently, however, AWSCHALOM[4] has reported the lower limits 2×10^{18} and 0.5×10^{18} years for the same isotopes in direct contradiction of the results presented by McCARTHY. This seems to indicate that McCARTHY's results must be interpretated as a queer statistical accident.

Finally, a measurement performed at Los Alamos[5] on Nd^{150} has also given a negative result. Gram quantities of the Nd^{150} isotope were spread in a 30 mg/cm^2 layer between very thin aluminium coated Mylar films, placed in a liquid scintillator and viewed by two 5-inch photomultiplier tubes which register coincidences and send their respective pulse heights to be added and the pulse sum analysed in a 100-channel analyser. This main system was entirely

[1] J. H. FREMLIN and M. C. WALTERS: Proc. Phys. Soc. Lond. A 65, 911 (1952). — M. I. KALKSTEIN and W. F. LIBBY: Phys. Rev. 85, 368 (1952). — J. S. LAWSON: Phys. Rev. 81, 299 (1951). — R. M. PEARCH and E. K. DARBY: Phys. Rev. 86, 1049 (1952). — R. G.WINTER: Phys. Rev. 99, 88 (1955). — C. A. LEVINE, A. GHIORSO and G. T. SEABORG: Phys. Rev. 77, 296 (1950).
[2] M. G. INGHRAM and J. H. REYNOLDS: Phys. Rev. 76, 1265 (1949).
[3] J. A. McCARTHY: Phys. Rev. 87, 194 (1952); 97, 1234 (1955).
[4] M. AWSCHALOM: Bull. Amer. Phys. Soc., Ser. II 1, 31 (1956).
[5] C. L. COWAN jr., F. B. HARRISON, L. M. LANGER and F. REINES: Private communication.

immersed in a 600 liter liquid scintillator viewed by twelve 5-inch photomultipliers and used as an anticoincidence shield against charged particle background. The results obtained with Nd[150] and Nd depleted to 0.064% gave no indication of a positive effect from reaction (27.1). The sensitivity was such that a half-life of 2×10^{18} years could have been easily detected, and the authors conclude that the assumptions underlying reaction (27.1) are not valid.

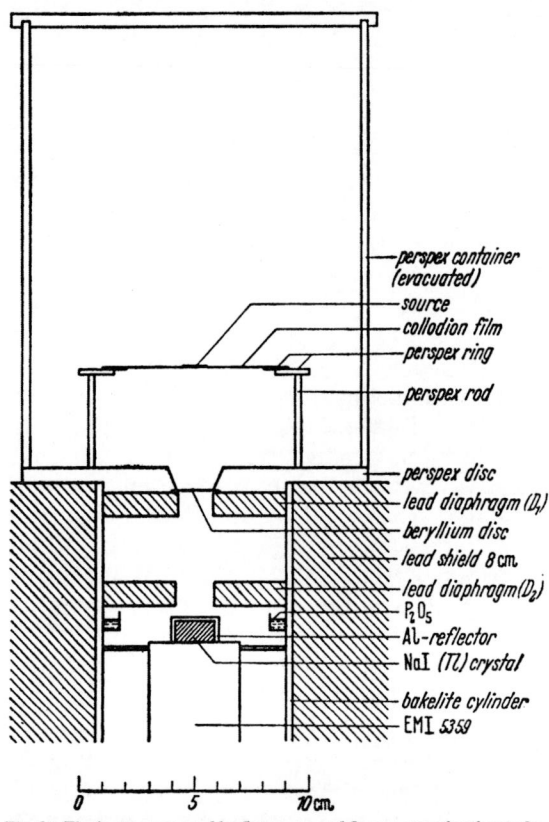

perspex container (evacuated)
source
collodion film
perspex ring

perspex rod

perspex disc
lead diaphragm (D₁)
beryllium disc
lead shield 8 cm.
lead diaphragm (D₂)
P₂O₅
Al-reflector
NaI (Tl) crystal
bakelite cylinder
EMI 5359

0   5   10 cm.

Fig. 89. The instrument used by STARFELT and SVANTESSON for the study of internal bremsstrahlung. (After N. STARFELT and N. L. SVANTESSON.)

28. Internal bremsstrahlung. α) Internal bremsstrahlung in β-decay. Internal bremsstrahlung is a second order process in which one interaction arises from β-decay and the other from electromagnetic interaction. The result of the reaction is the emission of an electron, a neutrino and a photon. The theory of the reaction has been described by several authors[1]. The transition is often called *radiative decay*.

As mentioned already in many other connections, the decay of RaE does not lead to excited nuclear states in the daughter and no nuclear γ-rays are emitted. ASTON[2] found, however, that weak electromagnetic radiation appears, and he showed that the X-ray energy equals approximately one percent of the energy carried by the β-rays. Furthermore, he suggested that the X-ray spectrum is continuous.

ASTON studied the X-rays in ionization chambers, and this procedure has since been applied in many investigations. However, the weak intensity of the X-rays necessitates the use of strong sources in such investigations and, consequently, the source becomes relatively thick, and the decay electrons are slowed down in the source material and give rise to *external bremsstrahlung*. The presence of external bremsstrahlung complicates the investigations, of course, and many of the older experiments may be considered in doubt because sufficient care was not taken to dispose of this effect. The usual procedure is to stop the electrons in various absorbers of different Z and extrapolate the results to $Z = 0$ where the external bremsstrahlung effect disappears.

Progress was really made when scintillation counters were introduced. The high efficiency of scintillation counters for X-rays permits the use of weaker sources and the above effects from external bremsstrahlung are therefore di-

[1] J. K. KNIPP and G. E. UHLENBECK: Physica, Haag 3, 425 (1936). — F. BLOCH: Phys. Rev. 50, 272 (1936). — C. S. W. CHANG and D. L. FALKOFF: Phys. Rev. 76, 365 (1949). — L. MADANSKY, F. LIPPS, P. BOLGIANO and T. H. BERLIN: Phys. Rev. 84, 596 (1951).
[2] G. H. ASTON: Proc. Cambridge Phil. Soc. 23, 935 (1927).

minished. Many recent experiments using scintillation detectors have been reported[1]. Great care has been taken to eliminate X-rays produced by the stopping of electrons in the source and the material surrounding the source. The instrument used by STARFELT and SVANTESSON is shown in Fig. 89. Light elements which give relatively little external bremsstrahlung are used wherever possible. The results of LIDÉN and STARFELT for internal bremsstrahlung from P^{32} are shown in Fig. 90. The experimental points are compared with the theory of KNIPP and UHLENBECK, which gives the solid line in Fig. 90. For low energies the agreement is good; for higher energies a deviation between theory and experiment appears.

NOVEY has also studied the angular correlation between electron and photon in radiative decay of RaE. His results are in good agreement with the theory, but his investigations do not extend to the high energies where the results of LIDÉN and STARFELT show a small discrepancy with theory.

β) Internal bremsstrahlung in electron capture. Internal bremsstrahlung in K capture occurs when a K electron is captured and a neutrino and a photon are emitted. The reaction is also often called radiative capture. The theory for this reaction has been developed by MORRISON and SCHIFF[2] and in more detail by GLAUBER and MARTIN[2].

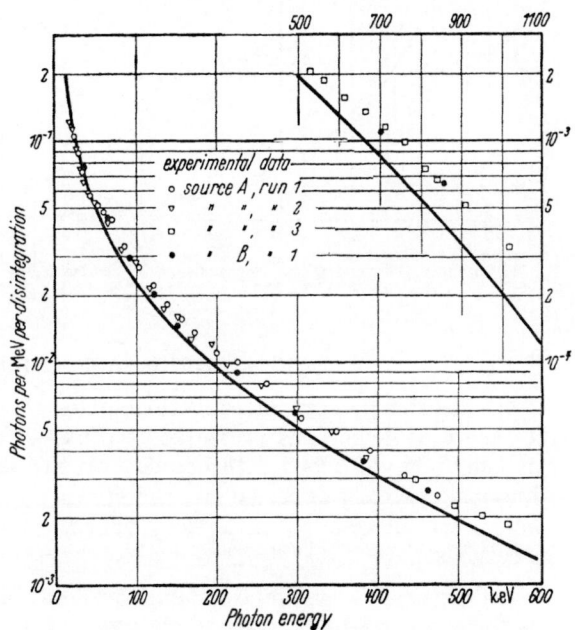

Fig. 90. Internal bremsstrahlung spectrum from P^{32} as obtained by LIDÉN and STARFELT. (After K. LIDÉN and N. STARFELT.)

Experimentally the reaction is often used to determine the energy available in K capture. MORRISON and SCHIFF neglected Coulomb effects and included $1S$ electrons only, whereas GLAUBER and MARTIN included Coulomb effects and considered other electrons in addition to the $1S$. The differences in their theoretical results appear mainly in the low energy region. The high energy region used in the determination of K capture energies is well represented by the simple distribution (26.2). It is seen from this equation that plotting $(P(k)/k)^{\frac{1}{2}}$ as a function of k, where P is the measured distribution, results in a straight line cutting the energy axis at $k=W_0^K$, i.e., the energy release in the K capture reaction. This bremsstrahlung plot is analogous to the Fermi plot for β-decay and is widely used for the determination of the energy release in K capture.

[1] L. MADANSKY and F. RASETTI: Phys. Rev. 83, 187 (1951). — P. BOLGIANO, L. MADANSKY and F. RASETTI: Phys. Rev. 89, 679 (1953). — T. B. NOVEY: Phys. Rev. 84, 145 (1951); 86, 619 (1952); 89, 672 (1953). — M. GOODRICH, J. S. LEVINGER and W. PAYNE: Phys. Rev. 91, 1225 (1953). — G. A. RENARD: J. Phys. Radium 14, 361 (1953). — K. LIDÉN and N. STARFELT: Phys. Rev. 97, 419 (1955). — N. STARFELT and N. L. SVANTESSON: Phys. Rev. 97, 708 (1955).

[2] P. MORRISON and L. I. SCHIFF: Phys. Rev. 58, 24 (1940). — R. J. GLAUBER and P. C. MARTIN: Phys. Rev. 95, 572 (1954).

The bremsstrahlung spectrum is more difficult to investigate in more detail. Geometry and efficiency corrections must be taken into account and scattering avoided. (In contrast to the case of β-decay no troubles arise from external bremsstrahlung.) Recently Lindqvist and Wu[1] have studied the radiative capture of A^{37} with great care. They obtained the spectrum shown in Fig. 91 and compared it with the two versions of the theory as shown. The theoretical curves have been corrected for the experimental effects. It is seen that the deviation between the theories and experiment occurs mainly at low energies. The experimental curve does not agree entirely with either of the two theoretical curves, but is such that the general shape at high energies is well represented by Eq. (26.2).

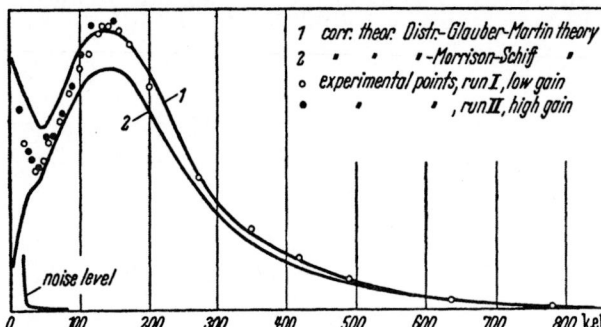

Fig. 91. The internal bremsstrahlung spectrum from A^{37} as obtained by Lindqvist and Wu. (After T. Lindqvist and C. S. Wu.)

Concluding remarks.

The field of β-decay investigations gives good support to the ideas behind the theory of β-decay as presented by Fermi. Most of the experimental facts can indeed be described satisfactorily by the theory. The single outstanding additional discovery is parity non-conservation as introduced by Lee and Yang. The present information on β-decay experiments seems to permit the following conclusions.

1. β-particles originate from nuclear decay and are electrons or positrons. β-particles are in a fully polarized state when emitted, that is the degree of polarization is v/c. The direction of the polarization is so that β^--particles are lefthanded (polarized antiparallel to the momentum) whereas β^+-particles are righthanded.

2. Neutrinos are neutral spin $\frac{1}{2}$ particles of vanishing rest mass. This is to be understood so that no evidence for a finite mass has been found. At least two different neutrinos exist: The fully polarized lefthanded antineutrino associated with β^--decay and the righthanded neutrino associated with β^+-decay.

3. All experimental evidence supports the view that the β-decay is a direct three body decay.

4. The electron capture appears as a two-body decay as expected from theory.

5. The β-decay follows the usual laws of conservation: energy, momentum and angular momentum. The interaction is invariant against time reversal and (what is in effect the same) a combined parity and charge conjugation transformation. The interaction violates the usual law of invariance against a parity or charge conjugation transformation. As a result of the time reversal invariance all coupling constants may be chosen real.

6. Spectral shapes and directional and polarization correlations agree within the experimental uncertainties with the predictions of the theory except for small deviations of the β-spectrum, which are not understood.

[1] T. Lindqvist and C. S. Wu: Phys. Rev. **100**, 145 (1955).

7. The G-T interaction is mainly axial vector. The Fermi interaction is mainly vector. The possible contributions of tensor (G-T) and scalar (Fermi) are presumably small although no very accurate statement is possible.

8. At present very little may be said about the possible role of the pseudo-scalar interaction.

9. The relative sign of Fermi and G-T coupling is negative.

10. The G-T and Fermi couplings contribute numerically by approximately equal amounts. The contribution of the G-T interaction is slightly larger than that of the Fermi interaction.

11. In cases where nuclear theory is sufficiently advanced to permit the calculation of the matrix elements the lifetimes involved in β-decay may be satisfactorily accounted for. In most cases, however, such calculations are not possible.

12. The β-decay experiments made until now are described with sufficient accuracy by means of a four component theory as well as by means of a two component theory for the neutrino.

13. The experimental study of mesonic effects in β-decay is at an early stage of development and no definitive results have been obtained.

These statements have been made from the point of view of establishing a scheme which summarizes the experimental results. From a theoretical point of view, however, it must be remembered that the coupling constants which have been disregarded above may possibly be as big as 10 to 20% of the main coupling constants. This fact is worth remembering when discussing the stringent applicability of e.g. the two-component theory or the universal Fermi coupling theories.

The material necessary for the establishment of these facts has been presented in the present article. Little emphasis has been placed on the historical development of the study of β-decay and detailed references to original articles discussed in references [1], [3], [4] and [5] are in general not included in the present chapter. This has been done in order not to introduce an overwhelming number of references. But by doing this the authors have had to omit names of many colleagues who have made important contributions to our knowledge of experimental material. For this fault the authors wish to offer their sincere apology.

General references.

[1] AJZENBERG-SELOVE, F., and T. LAURITSEN: Nuclear Phys. 11, 1 (1959). — American Institute of Physics Handbook. New York: McGraw-Hill 1961 (to appear).
[2] SIEGBAHN, K.: Beta and Gamma Ray Spectroscopy. Amsterdam: North-Holland Publishing Co. 1955.
[3] STROMINGER, D., J.M. HOLLANDER and G.T. SEABORG: Rev. Mod. Phys. 30, 585 (1958).
[4] PATTER, D.M. VAN, and W. WHALING: Rev. Mod. Phys. 26, 402 (1954); 29, 757 (1957).
[5] KING, R.W.: Rev. Mod. Phys. 26, 327 (1954). — L.J. LIDOFSKY: Rev. Mod. Phys. 29, 773 (1957).

Sachverzeichnis.

(Deutsch-Englisch.)

Bei gleicher Schreibweise in beiden Sprachen sind die Stichwörter nur einmal aufgeführt.

Subject Index.

(English-German.)

Where English and German spelling of a word is identical the German version is omitted.